GERMAN DAGGERS OF WORLD WAR II

German Daggers

OF WORLD WAR II

A Photographic Reference

VOLUME FOUR

- Recently Surfaced, Rare and Unusual Daggers
- Hermann Göring Bejeweled Dress Daggers
- Reproductions
- Solingen Update

Thomas M. Johnson

Schiffer Military History
Atglen, PA

ACKNOWLEDGEMENTS

This new reference book would not have been possible without the kind contributions of the following friends, collectors, and historians – Larry Allen, Kim D. Alstott, Edward A. Anderson, Jack Arnold, H.K. Askildsen, the late James P. Atwood, Rheinard Bauernfeind (Austria), Michael Bauman, Peter Beltran, R. James Bender, Jacques Bliard (France), Björn Blomgren (Sweden), Larry Bolton, Phillip Borchert, Malcolm Bowers (United Kingdom), Jeff Bradford, Barry Brown (United Kingdom), Doug Brown, Larry Browning, Jean-Jacques Buigne (France), Jason P. Burmeister, Dr. Paul Bushman, George Canady, Edward J. Carbone, Paul Carroll, Frances Catella (France), David Chvatal, David Cowart, TSGT (Ret.) T.W. Cunningham, Ben Curtis, Frank D'Agostino, Gailen David, Joel Day, Edward L. Diefenthal, David Delich, Erika Diehl, Ruhrik Diehl, Victor Diehl, Tom Donnell, Charles Eliopoulos, Colonel Michael Everett, Leo Fackler, the late Fred Fallgren, Vinnie Finelli, the late W. E. Fitzgerald, the late Anthony T. Flotte, Raul Forte (Argentina), Robert Frazier, the late George R. Fuerst, Michael Galati, Sterling ("Chip") Gambino, Michael Gantt, Uwe Gartner (Germany), Peter Gauf (Germany), J. Jay Gillespie, Dennis Gogolski, Craig Gottlieb, Doug Gow (New Zealand), Albert ("Skipper") Greenwade, Claude Groell (France), Dr. Barbara Grotkamp-Schepers (Germany), Jill Halcomb, Peter F. Hammond, N. Neil Hardin, Lothar Hartung (Germany), Michael J. Hardung, Ian Hargreaves (Canada), Bud Hasher, Tony J. Hedges (Australia), N.C. ("Dutch") Heilman, Gordon Henger, Joseph H. Hess, Richard Hill, Robyn Ann Hoffmaster, Thomas D. Howard, Robert Hritz, Jean-Marie Irlande (France), Charles H. Jenkins, III, Carter Jennette, Robert Johns, Edwin Johnson, Gary Johnson, the late Ed Kailen, Ralph Kaschai, Lemmy Kilmister, William Klotz, Kai Knehr (Germany), John Kopik, Wolfgang Kriechhammer (Austria), Bavin J. Lane (United Kingdom), William Laret (United Kingdom), Paul Lee (United Kingdom), Edmund F. Leland, George Lepre, James Lesher, Ron Levin, Leif Linde (Sweden), LTC (Ret.) Michael E. Little, Harry Logan, Jess Lukens, Manion's International Auction House, Charles E. Marx, Robert McCarthy, David A. McGlothlin, Robert Meistrell, James Mendelsohn, Charles Mermet, Arthur "Addi" Meyer (Germany), the late Dr. Julian Milestone, Michael A. Moeller, Andrew Mollo (France), Brian Molloy (UK), J. A. Morrison, Robert D. Moser, Robert P. Moses, Dan Nagle, Russ Neighbors, Earl Ohlgren, Aurelius Pajkert, George E. Patient, Darko Pavlovi (Croatia), John P. Pearson, John E. Pepera, Robert J. Petrik, Larry Pistole, Lloyd C. Polk, Joseph Polsinelli, Philip Poole, Thomas W. Pooler, Adam Portugal (Germany), Dr. Robert Procaccini, Frank F. Quinn, Mark Ready, Larry Redgate, Ren Reynolds, Carl Righter, Harvey Riley (England), Denny Roach, David Rome, SGM (Ret.) Kenneth Rouse, W.P.B.R. Saris (Holland), Chuck Scaglione, John G. Schleicher, Erwin Schulz (Germany), Robert Sevier, the late Michael Shaughnessy, William (Bill) Shea, Andy Shoredits (South Africa), John Sieber, Kris Simoens (Belgium), MAJ (Ret.) Charles E. Snyder, Jr., the late Phil Snyder, Andy Southard, Jr., Jack Staehle, T. S. Stafford (United Kingdom), Dr. Ernst Stangen (Germany), the late Roger Steele, Christopher Stein, Frederick J. Stephens (United Kingdom), Martin Stiles (United Kingdom), Joe Stone, the late Ben Swearingen, Art Sylvie, Alan Taugourdeau (France), Dr. Fred Thrash, Jim Toncar, George Tritton (England), Bryan Turner, Richard "Dick" Van Heuvelen, Kurt Van Laere (Belgium), Odd Vehusheia (Norway), Donald Vernon, Roger Viollet (France), the late Robert Waitts, Fritz Waltermann (Germany), Tony Ward (United Kingdom), Ron Weinand, Curt Weskalnies, Martin White, Keith Wilson, Jr., Stewart Wilson (United Kingdom), Dr. William A. Windrum (Canada), Tom Winter, Michael Wipf, Thomas T. Wittmann, Stephen D. Wolfe, Kevin Woods, Ron Woodward, Frank Worley, Robert Zackman, and David Zeller. Without any one of the aforementioned persons, this reference would not have been as complete or accurate. The importance of each of their contributions to this work is greater than I can adequately express. To those whose contributions I have neglected to note here from sheer failure of memory, please accept my sincere apology. One final note, I would like to especially thank my wife Tink and my personal secretaries Donna Kines and Reva Darlene Fox for their continued and unwavering support for the past two decades.

Book design by Robert Biondi.

Library of Congress Catalog Number: 2004116057.

Printed in China.
ISBN: 0-7643-2206-0

We are always looking for people to write books on new and related subjects. If you have an idea for a book, please contact us at the address below.

Published by Schiffer Publishing Ltd.
4880 Lower Valley Road
Atglen, PA 19310
Phone: (610) 593-1777
FAX: (610) 593-2002
E-mail: Info@schifferbooks.com.
Visit our web site at: www.schifferbooks.com
Please write for a free catalog.
This book may be purchased from the publisher.
Please include $3.95 postage.
Try your bookstore first.

In Europe, Schiffer books are distributed by:
Bushwood Books
6 Marksbury Ave.
Kew Gardens
Surrey TW9 4JF
England
Phone: 44 (0)20 8392-8585
FAX: 44 (0)20 8392-9876
E-mail: Bushwd@aol.com.
Free postage in the UK. Europe: air mail at cost.
Try your bookstore first.

Contents

German Daggers of World War II

CHAPTER TWENTY-THREE

Recently Surfaced Rare and Unusual Dress Daggers of the Third Reich

"If I have ever made any valuable discoveries, it has been owing more to patient attention, than to any other talent." — Isaac Newton

Previously Unidentified Third Reich Dagger Model Discovery!

After all of the numerous contemporary reference books on the specific subject of German Third Reich dress daggers, one might assume that the discovery of a previously unknown standard dagger model from the Third Reich era was well past. However, nothing could be further from the truth, because early last year, a previously totally ignored and unmentioned model has come to light. What are the odds of discovering as late as the year 2003 a previously unknown standard Third Reich dress dagger? None of the many comprehensive reference books cited above, including my own numerous volumes on the subject, even make mention of the dagger model described below.

This ultra-rare discovery came into the possession of the author during May 2003. The cased dagger complete with unique portepee was privately purchased from the grandson of the original owner, Julius Grün, in Köln, Germany.

Since the dagger itself appeared to be practically identical to a 1st Model Navy dress dagger, it posed a dilemma since the grandson was confident that his grandfather had never served in the German Navy or elsewhere in the German *Wehrmacht*.

What was known about this interesting and unusual dress dagger were the following few facts:

1.) the cased presentation dagger had been purchased from Julius Grün's own grandson.

2.) the cased dagger had been in the sole possession of the family for three generations.

3.) according to the grandson, his grandfather had never served in the German Navy.

4.) the grandson stated that his grandfather was an early member of the German Auto Club and had personally been selected to escort the body of World War I Flying Ace Baron Manfred von Richthofen ("the Red Baron") from France back to Germany for burial. It should be noted that this claim has not been verified by the author to date.

5.) the grandson also provided a period professional photograph of his grandfather behind the wheel of a large early Mercedes automobile, as well as several other German Auto Club cased pre- and post-war decorations and an .800 fine silver auto or sleeve badge with his grandfather's full name on the reverse side.

The unattributed cased dagger had been in the U.S. less than seventy-two hours when well-known militaria dealer/collector Ronald Weinand of Quincy, Illinois, visited the Johnson Reference Books and Militaria office complex in Fredericksburg. Ron inquired about what rare pieces had been found in Europe on my last trip, and when the cased unattributed dagger was placed in front of him, he not only was duly impressed (an understatement if there ever was one!), but after much thought, offered the theory that this presently unattributed dagger had to be a Third Reich Auto Club dress dagger.

To date, the collecting community was only aware of an *Imperial* Auto Corps dress dagger. It is interesting to note that a dagger of the existing Imperial Naval Officer configuration was selected and authorized for wear by the senior officials of the civilian para-military organization, the *Freiwilliges* (Volunteer) Automobile Corps.

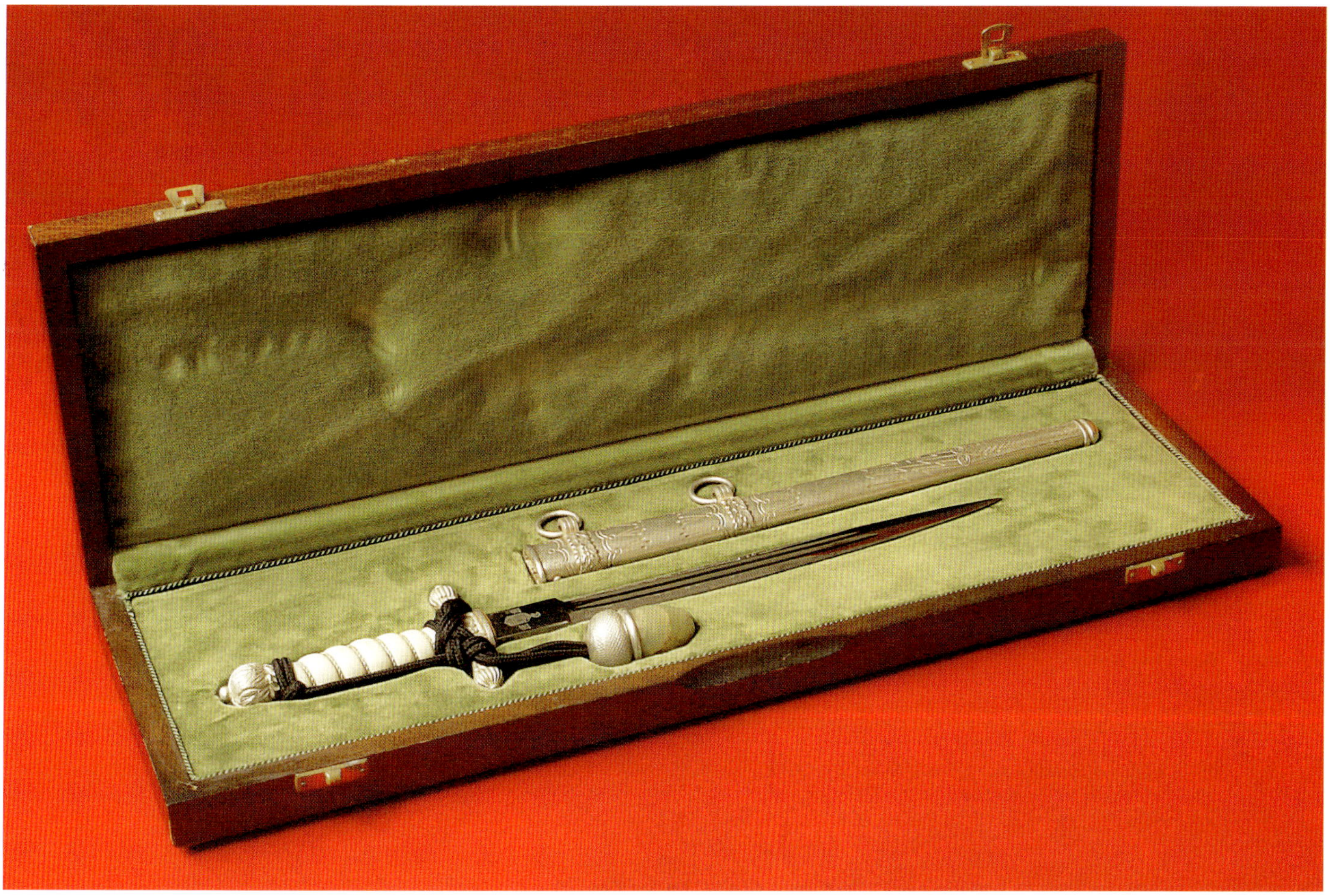

The Julius Grün Third Reich Auto Club dagger is pictured in its custom-made recessed presentation walnut case. The interior lid is covered in pale sage green velvet matching the color of the onyx or soapstone acorn with the lower fitted portion covered in the same material and fitted to display the dagger and scabbard. A truly unique presentation Third Reich dagger and portepee. *Author's collection. Photo by Charles H. Jenkins, III.*

The Julius Grün Third Reich Auto Club dagger is complete with a custom-made dark brown walnut grained presentation case with a silver oval emblem of the *ADAC* with Imperial eagle and crown inset into the exterior lid. *Author's collection. Photo by Charles H. Jenkins, III.*

RIGHT: The custom-made dark brown walnut grained presentation case housing the Third Reich Auto Corps dress dagger features a silver oval emblem of the *ADAC* with Imperial eagle and crown inset into the exterior lid. *Author's collection. Photo by the author.*

BELOW: The Julius Grün Third Reich Auto Club dagger is pictured displayed on top of its walnut grained presentation case. Note the integral scabbard bands (see the last few paragraphs of this article). *Author's collection. Photo by Charles H. Jenkins, III.*

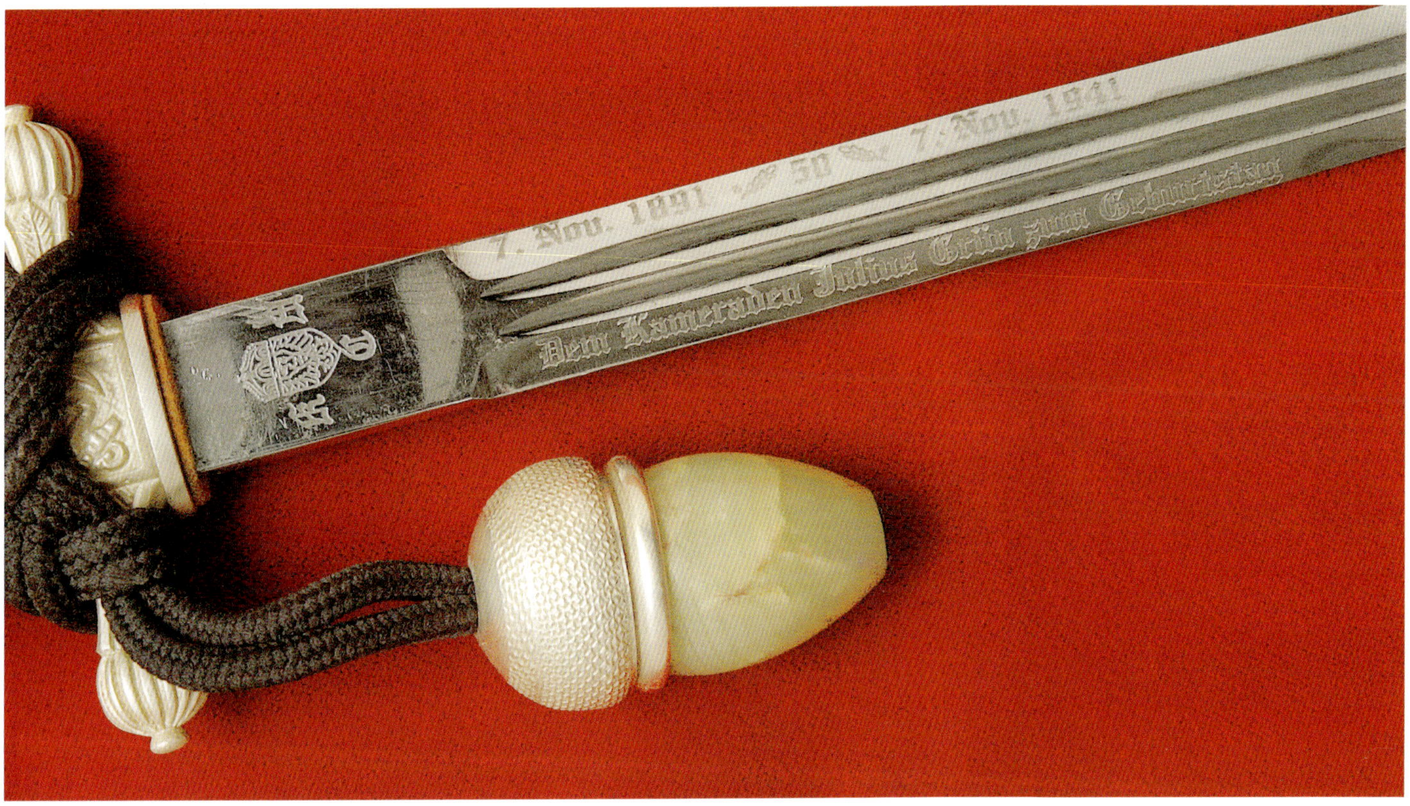

The obverse ricasso is etched with the crown over Imperial eagle emblem with the *"KAC"* initials of the Imperial German Automobile Club. The obverse top of the blade is etched, "7.Nov.1891" and, "7.Nov.1941" flanking an oak leaf and acorn motif on either side of the number "50" (Julius Grün's fiftieth birthday). Again, note the highly unusual portepee consisting of a black woven cord tied in a Naval reef knot complete with acorn consisting of a hand-stippled silver crown atop a green onyx or soapstone acorn. *Author's collection. Photo by the author.*

Clearly visible on the reverse blade ricasso is the late over-the-shoulder Carl Eickhorn trademark. As mentioned in the text, this particular Eickhorn trademark is associated with the "late 1941 to discontinuation of dagger production in 1944." The Grün dagger was presented on 7 November 1941 and is consistent with the Eickhorn trademark dating. *Author's collection. Photo by the author.*

The Julius Grün presentation Third Reich Automobile Club dagger is pictured atop the 1939 *DDAC* roadmap printed in Munich. Note the eagle/swastika *DDAC* logo on this civilian road map. *Author's collection. Photo by the author.*

The Automobile Corps

The automobile had scarcely been in production when the German Army developed a strong interest in the use of motor vehicles for military purposes. Military vehicles were first employed in an experimental role during the *Kaisermaneuver* of 1899. The first motor transport elements were introduced into the German Army in 1906. This Corps expanded with the growth of the German transportation industry and had achieved a high degree of development by 1914. At the beginning of the war the military motor vehicle was considered by many professional soldiers to be only a passing folly. However, by the end of World War I motorized transportation had established itself as a necessity beyond any doubt.

During peacetime, the transportation of the German Army was the responsibility of the *Kraftfahrttruppen*. A Bavarian unit was organized in 1908, and a Prussian Battalion consisting of four companies was developed in 1911. To augment the regular transportation specialists, the Army was allowed to call upon the members of the Volunteer Automobile Corps. This civilian organization was founded at the turn of the century to provide the Army with skilled drivers. As a basis for membership, each member must own his own automobile and be ready to offer both his car and his services to the state in event of the outbreak of hostilities.

As an act of patriotism, a number of Germans with automobiles joined the Volunteer Automobile Corps. Many members of the nobility, as well as officers in the Army, maintained membership in the Corps. A special uniform *including a dress dagger* was developed to be worn during the many parades and formal activities in which Corps members participated.

The Prussian Automobile Corps Dagger

The Prussian Automobile Corps dagger was basically a silver version of the long model 1901 Imperial Naval Officer's dagger. The hilt contained a silver open crown pommel identical to its Naval counterpart. The grip of most Senior Official Patterns was made of genuine ivory. Double-spiral grooves ran from the lower left to the upper right. The number of grooves was generally six, and the grip was double-wrapped with either silver or gold double-twist grip wire.

The crossguard was identical to the 1848 Pattern Navy with the exception of the design motif of the obverse rectangular quillon block. This block contained an affixed brass, gold plated hunting horn, which faced in either direction. The crossguard reverse on most examples contained a push button used to compress the scabbard retaining spring. An early pattern was produced which used a piece of bowed tension steel as the retaining spring. This configuration did not require the quillon block push button. The metallic composition of the pommel and crossguard was either brass, aluminum or steel. Aluminum and brass hilts generally had a silver wash, while those of steel were usually nickel-plated.

The Automobile Corps dagger blade was based on the 1890 Naval Cadet Pattern. The nickel-plated blades were of the double-edged stiletto variety and were either plain or etched. The blade ricasso was plain or maker-marked. Blades have been observed marked with the knighthead of WK&C, the back-to-back squirrels of Carl Eickhorn, "Siegfried" of Pack, Ohliger and Co. and the crowned "Z" of Clemen und Jung.

The Automobile Corps scabbard was either dark brown or black leather with silver fittings. The scabbard fittings may be solid German silver, silver plated brass, or nickel-plated steel. The top scabbard fitting contained a frog stud while the lower fitting was scalloped and engraved to match the top fitting and ended in a ball-like tip.

The dagger was worn on the left side and suspended from the belt in either a black or brown leather carrying frog. The front of the frog was V-shaped leaving the top of the scabbard fitting exposed.

A special knot was produced for wear with the Automobile Corps dagger. The silver bullion portepee strap was 7 millimeters wide and contained a centrally located red stripe with a dark blue stripe on each side of the red one. The silver slide and stem were interwoven with red and blue silk, and the acorn ball was silver and contained a blue velvet core.

The Bavarian Automobile Corps Dagger

A distinctive Automobile Corps dagger was worn by members of this Bavarian organization. The dagger was basically of the same configuration as its Prussian counterpart, but with certain conspicuous characteristics.

The Bavarian pommel is easily recognized. It is composed of eight arches, which extend from the base and meet at the top with a traditional orb and cross. The arches were decorated with raised circles enclosed by raised edges along their entire length. The pommel was either silver, gilt or painted black with the raised designs touched in gold. The pommel screwed on the blade tang and was removable.

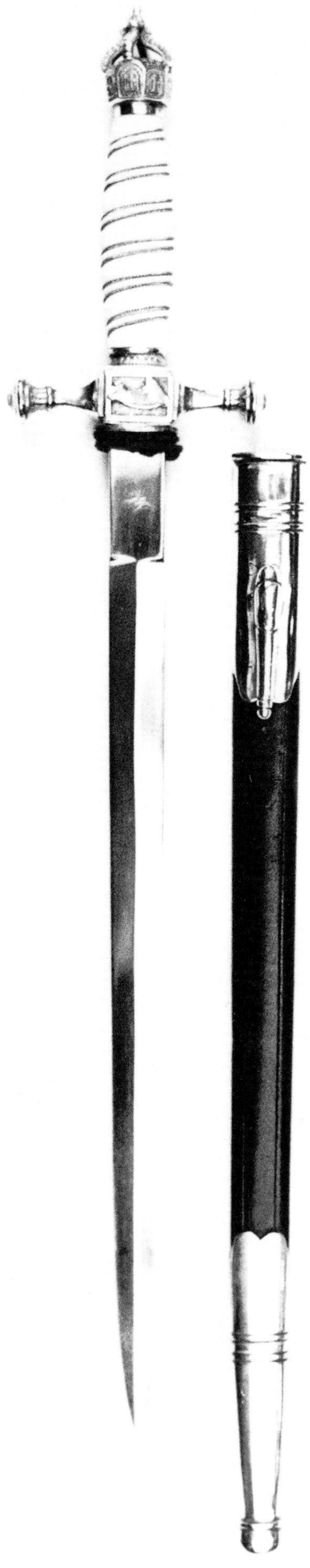

Prussian Auto Corps dagger. This model features the stamped and engraved, rather than cast crossguard horn insignia. This specimen also features a silver grip wrap. The variation frog stud is easily recognized. The blade is stamped with the knight's head of WK&C. *Erika Diehl collection. Photo by Victor Diehl.*

The grip was genuine ivory with a double or single wrap of twisted gold or silver wire. The organizational insignia on the obverse crossguard was the same as on the Prussian model. However, Bavarian models have also been observed with no insignia attached to the crossguard.

The metallic composition of the hilt fittings was either brass, aluminum or steel. The aluminum and steel fittings were usually silver plated or black enameled, while the brass fittings were either silver plated, gilted or lacquered. Bavarian Auto Corps blades are found in the same variations of size and etching as the Prussian variety.

The scabbard body and fittings for the Bavarian Auto Corps dagger displayed the same basic characteristics as the Prussian model, with the exception that some models had steel scabbard fittings, which were painted with black enamel to match the hilt of the dagger. The dagger was worn suspended in the same manner as the Prussian model.

The reader can find a complete, detailed description of these early Imperial Auto Corps daggers on pages 240-247 of *Collecting the Edged Weapons of Imperial Germany, Volume I* by the author and Thomas T. Wittmann. However, no original 1933-1945 Third Reich era Solingen sales catalog or contemporary reference mentioning a Third Reich Auto Corps dress dagger is known to the writer of this article.

After their seizure of power in 1933, the National Socialists felt that the duplication of several automobile and motorcycle clubs (*ADAC, AvD, DTC*, and *NDA*) was a definite thorn in the side. Under pressure from the *NSDAP* the representatives of these four major clubs signed on 24 July 1933 a paper that entered into motoring history as the "Pact of Baden-Baden." The leadership of the combined German motoring organization was immediately transferred to the National Socialist Motor Corps (*NSKK*). Finally, on 27 September 1933 the united club was inaugurated with the name/acronym *DDAC* (the German Automobile Club). This provides for the history of the follow-on *ADAC* organization. As international travelers are well aware, the German *ADAC* organization remains a visible major civilian organization (similar to the AAA in the U.S.) to the present date.

Cased Presentation Third Reich Automobile Club Dagger, Medal/Document/Photograph Grouping Belonging to Julius Grün

This ultra rare grouping was recently obtained in Germany from the grandson of the original owner, *Herr* Julius Grün, a member of the *Kaiserlicher Automobil-Club (KAC)* (Imperial Automobile Club) and the post-Imperial *Allgemeiner Deutscher Automobil-Club (ADAC)* (General/Universal German Automobile Club). The grandson stated that his grandfather was in some way, possibly in his position as a senior member of the Civilian Automobile Club,

Bavarian Auto Corps dagger. This dagger has aluminum hilt fittings painted black with gold trim. The grip is wrapped with gold wire. The crossguard horn is cast brass and gilted. All hilt fittings as well as the blade tang are numbered, "6." The floral pattern blade etch is very distinctive. The scabbard fittings are steel, painted black. *Rurik Diehl collection. Photo by Victor Diehl.*

involved with the return from France to Germany in 1926 of the body of World War I aerial ace Baron Manfred von Richthofen (the Red Baron). The grouping consists of the following items:

A. Ultra Rare Cased Presentation Third Reich Automobile Club Dagger. Carl Eickhorn, over-the-shoulder squirrel TM.[1] This superb dagger is in the form of a silver plated model 1929 German Naval dagger with a unique custom-made portepee. Apparently, the Weimar and Third Reich era Automobile Clubs carried on the Imperial tradition of using a silver version of the German Naval dagger for the Club edged weapon. This particular dagger was presented to *Herr* Grün as a 50th birthday present on 7 November 1941 by his Automobile Club comrades from *Gau Köln*. Silver plated hilt fittings show only faint wear to the highlights with 100% of the frosted silver remaining in the recesses of the flaming ball/wave type pommel and crossguard with fouled anchor design. A white celluloid-over-wood grip shows virtually no wear with no cracks or chips and is complete with tight twisted silver wire wrap. The obverse ricasso is etched with the crown over Imperial eagle emblem with the initials, *"KAC"* of the Imperial Automobile Club. The obverse top edge of the blade is etched, *"7.Nov. 1891"* and *"7.Nov. 1941"* flanking an oak leaf and acorn motif on either side of the number "50". The bottom obverse edge of the blade is jeweler engraved in gothic letters, *"Dem Kameraden Julius Grün zum Geburtstag"* (From the comrades of Julius Grün on his birthday). The reverse top edge of the blade is jeweler engraved in gothic letters, *"Die Kraftfahrer des KAC Gau Köln"* (the driver/motorist of the Imperial Automobile Club *Gau Köln*). The dent-free silver plated lightning bolt scabbard shows virtually no wear with 100% of the factory frosting remaining throughout the scabbard, oak leaf suspension bands, and suspension rings. No discussion of this interesting new discovery would be complete without mentioning the *unique* portepee consisting of a black woven cord tied in a Naval reef knot complete with acorn consisting of a hand-stippled silver crown atop a pale green onyx or soapstone acorn. Knowledge of other known examples of the dagger and/or the portepee would be welcomed by the author, as this, obviously, would not be a one off Third Reich dagger. Dagger is complete with a custom-made dark brown walnut grained presentation case with a silver oval emblem of the *ADAC* with Imperial eagle & crown inset into the exterior lid. The interior lid is covered in pale sage green velvet matching the color of the onyx or soapstone acorn with the lower fitted portion covered in the same material and fitted to display the dagger and scabbard. A truly unique presentation Third Reich Automobile Club dagger and portepee.

B. Cased Automobile Club Sport Badges/Stickpins.
1. Cased ADAC Sport Badge in Gold w/Stickpin. Gilt badge features an eagle within an oak and laurel leaf wreath flanked by *ADAC*. Stickpin features the same design as the badge. Reverse badge features a needle-style pin and is maker marked, *"Weidmann."* Badge and stickpin are complete with unmarked burgundy leatherette hinged case with fitted interior.

2. Cased 1934 DDAC Stickpin. Stickpin features a black/white/red enamel emblem with eagle & swastika above an oak leaf sprig and the date "1934". Stickpin is complete with serrated pin and blue leatherette case.

3. Post-1945 Sport Badge Grouping. This grouping consists of two gold badges (one with stickpin) and one silver badge. Badges feature a female athlete holding aloft victory wreathes and flanked by the *ADAC*/eagle emblem.

C. Cased Automobile Club (KAC) Silver Badge and Mercedes Stickpin. Curved, oval, .800 marked silver badge measures 3.1875" long, by 2.75" wide and features a laurel leaf border surrounding the Imperial crown/eagle *KAC* emblem flanked by raised oak leaves above the raised inscription, *"Fahrer des Kaiserlichen Automobil-Clubs"* (Driver, the Imperial Automobile Clubs). The reverse is marked with the crescent, 800 and crown silver mark and engraved in gothic letters, "Julius Grün". Badge is complete with two attaching pins that are secured to the badge, with silver straps marked, *"A. Stübbe Berlin"*. Badge was most probably designed to be pinned on the upper arm of the driver's jacket to denote his duty position. Silver stickpin features the Mercedes circle star radiator emblem and is complete with serrated pin. Badge and stickpin are complete with a blue leatherette hinged case with lower fitted section.

D. Photograph and Roadmap Grouping.
1. Black & white 6.875"x9.25" Photograph. Shows *Herr* Grün behind the wheel of an early Mercedes 2-door Roadster. He is wearing a leather driving helmet and heavy coat, and the radiator/hood of the automobile is decorated with a victory wreath from a just-completed road race.

2. 1939 DDAC Roadmap Published in Munich. Depicts the *Deutschen Alpenlands* to include Munich, Innsbrück, Berchtesgaden, Salzberg and Vienna. Map measures 8.5"x40" when unfolded, and features multi-color terrain and road features. Interesting period map, complete with eagle/swastika DDAC logo.

What could be more exciting to a serious German edged weapon aficionado than the monumental discovery of a completely unknown/unrecognized standard Third Reich dagger model? The contents of this article begs the question – could many of the silver washed daggers attributed as Third Reich Navy Administrative dress daggers actually be Third Reich Auto Corps daggers? Perhaps the only true Navy Administrative daggers are the standard brass/gold washed daggers with hangers bearing aluminum or silver hardware. We know that the known Navy Administrative dress dagger that advanced collector SGM (Ret.) Kenneth Rouse traced back to the original owner's son in Germany was, in fact, a gilt (not silver washed) Navy dagger with hangers bearing white aluminum hardware (*see Appendix 3, pages 344-349 of Volume VIII, "Collecting the Edged Weapons of the Third Reich"*).

It now stands to reason that the silver washed Navy daggers that we all categorized as Navy Administrative daggers are, in fact,

Third Reich Auto Club sidearms, i.e. in order to forego the expense and the necessary time delay in designing an entirely new model dress dagger, the German Prussian and Bavarian Auto Clubs opted to adopt an Imperial Navy dagger design as their standard sidearm. It is only logical that the Auto Club during the Third Reich era would look once again at the standard Navy dagger design. Yes, the writer has come to the conclusion that what Ronald Weinand proposed as a theory has to, indeed, be a fact!

If this single dress dagger had not already done enough to advance our collective knowledge in this ever-popular hobby, another common belief that is completely negated by this interesting Third Reich dagger centers around the two scabbard bands. For many years, the knowledgeable dealers in the edged weapons field and the authors of contemporary reference books have purported the *integral* scabbard bands on German Navy daggers to be spurious. The common phraseology that is found in contemporary, widely accepted reference books reads like this, "the scabbard bands on some reproductions are made as an integral part of the scabbard as a single mould; original Solingen made scabbards were constructed with a one-piece ring scabbard band which was first heated then sweated onto a complete scabbard, when the band cooled it shrunk into place on the scabbard."

This 1944 German Automobile Club dagger would tend to totally disprove this long-standing theory. It now would appear possible that many of the 2nd Model Navy daggers with *integral* scabbard bands that have been labeled fakes for years are possibly very late production original pieces, i.e. since the obviously original Grün silver scabbard does feature the *integral* scabbard bands, it is only logical that the standard brass 2nd Model Navy scabbards that were manufactured in Solingen during the same time period (especially by the Carl Eickhorn firm) were most likely manufactured in the identical manner.

Fellow collectors, it is astonishing how much knowledge can be gleaned from a detailed, proactive examination of a single dress dagger. Happy collecting!

The Ultimate Hitler Youth *Fahrtenmesser*

Unquestionably, the finest single Hitler Youth (HJ) knife that this author has personally observed is the one featured in this article for "Militaria International" commemorating the 800th Year Anniversary of the *Burg Schloss* (Burg Castle) located just outside the German "Blade City" of Solingen and being shown for the first time to the collecting community. Not only is an *HJ Fahrtenmesser mit Damast Klinge* (HJ knife with Damascus steel blade) practically unheard of, the quality and singular beauty of this remarkable blade must be seen, to be appreciated.

How rare you ask? I can recall during the mid-1960s a serious HJ edged weapon collector in Missouri who had a large collection of assorted original HJ edged weapons consisting of at least twenty-five or thirty completely different variations of this standard model Third Reich sidearm, but none featuring a genuine Damascus steel blade. As a matter of fact, he had nothing in his entire collection that could begin to measure up to the unique HJ knife featured in this article.

Towering above the trees and strategically positioned atop a high mountain ridge just outside the city limits of Solingen and overlooking the well-known Wupper River is the centuries old *Schloss Burg* and the surrounding community of Solingen-Burg. A look at a detailed Solingen map will reveal that there are actually *two* Burgs, *Oberburg* and *Unterburg* with the community being divided between a portion of the population and businesses surrounding the castle on top and the remainder of the homes and businesses at the base of the mountain with nothing in between but a beautiful forest. Both communities are extremely picturesque, and each can claim a first class, well-known restaurant – *Gasthaus An Der Strassen Anno 1673* (the Guest House on the Streets Established 1673) at *Oberburg* and *Haus Striepen* at *Unterburg*. Actually, *Unterburg* reminds one of Tennessee's famous Gatlinburg resort in the heart of the Great Smokey Mountains, but is much smaller. The population of the two Burgs combined is only a few hundred persons. The two small communities are connected by a road, and by a commercial *Seilbahn* (a cable chair lift similar to those seen at ski resorts throughout Europe).

A visit to the Burg Castle and the surrounding grounds immediately carries one back several centuries to an enchanted land of knights and chivalry. The castle was designed as a fortress for the Counts of Berg (note the slightly different spelling from the surrounding village and is not a misprint) and served as a castle for the Dukes of Jülich-Berg-Klive-Mark.

Definitely, the most important single room within the castle walls of Burg Castle is the huge Knights' Hall. It served as the audience room and the gathering place of the knights. The Counts of Berg possessed, without a doubt, an especially splendid Knights' Hall. During the Middle Ages the effect of such a large room, which measures seventy-five feet long by thirty-six feet wide with a high ceiling, must have been overwhelming. The ordinary mortal during the time inhabited a very small house. When one entered so mighty a castle for the first time, he must have been astonished, or even bewildered.

During the thirteenth century the castle was occupied by Engelbert II, the Archbishop of Cologne and, therefore, the Duke of Westphalia, and was, reputedly, "the mightiest man north of the Alps". A large statue of Count Engelbert II stands in front of the castle and is pictured in this article. Although Engelbert only ruled for a very short reign, 1218 to 1225, he built the Burg Castle into an insurmountable mighty fortification. During November of 1225, Engelbert was murdered by his own cousin off the castle grounds. The cousin, Friedrich of Isenburg, was subsequently tortured to death in *Köln* (Cologne). The history and lore of the castle is rich with the Counts of Berg riding off on horseback to travel some 6,000 kilometers (3,750 miles) south to Jerusalem to participate in the Crusades, but the subject of this short article is the magnificent Hitler Youth sidearm, which recently surfaced in the vicinity of Solingen, and not the castle and its inhabitants.

The Hitler Youth Knife (*Fahrtenmesser*) was the first of the official *Nazi* daggers to be manufactured in quantity. It was designed for use as a scouting knife; however, it had the additional purpose of teaching the German youths the responsibilities of carrying arms or weapons at an early age. It was forbidden to wear the knife except when wearing the official HJ Uniform.[2]

In 1933 there were 2,100,000 members of the Hitler Youth movement. At this time only 70,000 of these youths, or only three percent of the total number of members, owned the Hitler Youth knife.[3] The ultimate goal of practically every Hitler Youth was the eventual acceptance into the SS.

In 1937 a special HJ *Führerdolch* (Leader's dagger) was introduced. Prior to this, the HJ leaders had worn the same *Fahrtenmesser* as the young boys. Some pieces were more elaborate than others, incorporating staghorn grips and ornately etched hilts and blades, depending upon the amount of money each individual

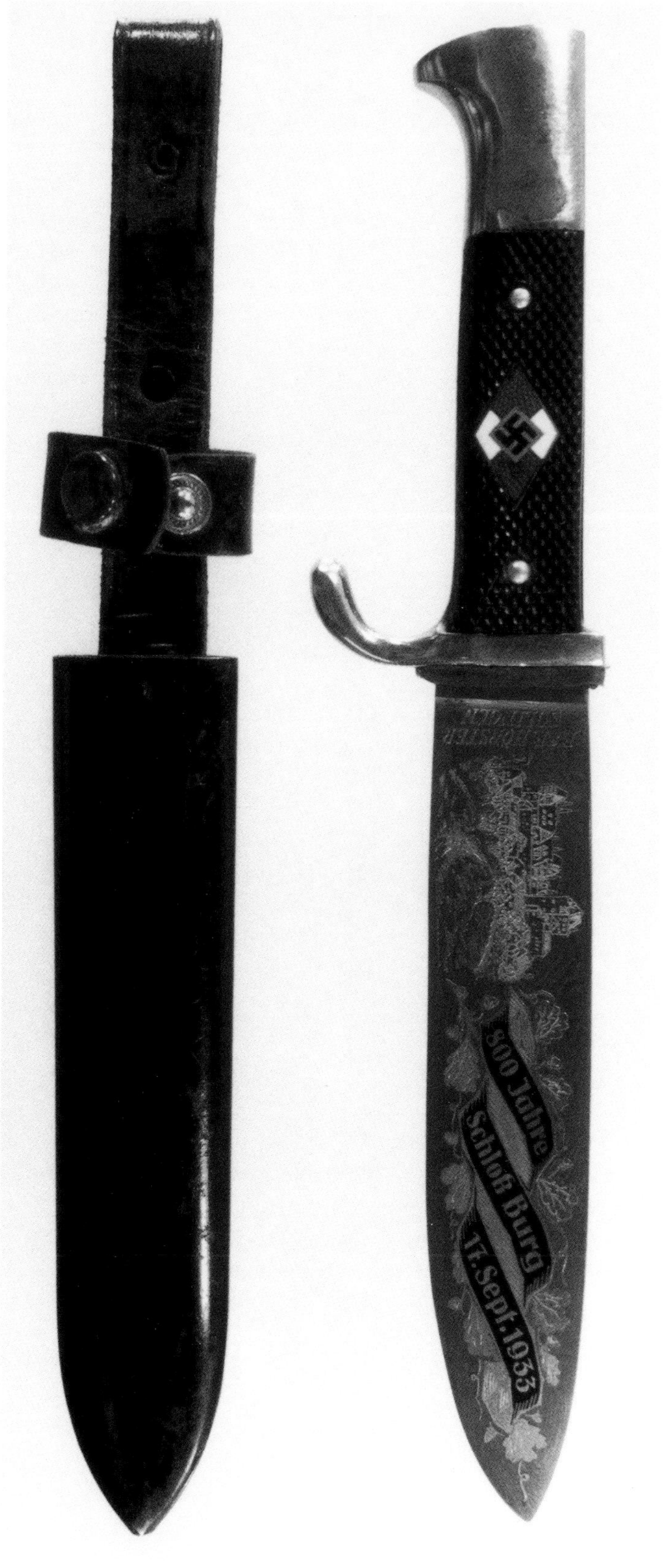

The one-of-a-kind Burg Castle Hitler Youth (HJ) knife is pictured out of the scabbard – obverse full-length view. *Author's collection. Photo by the author.*

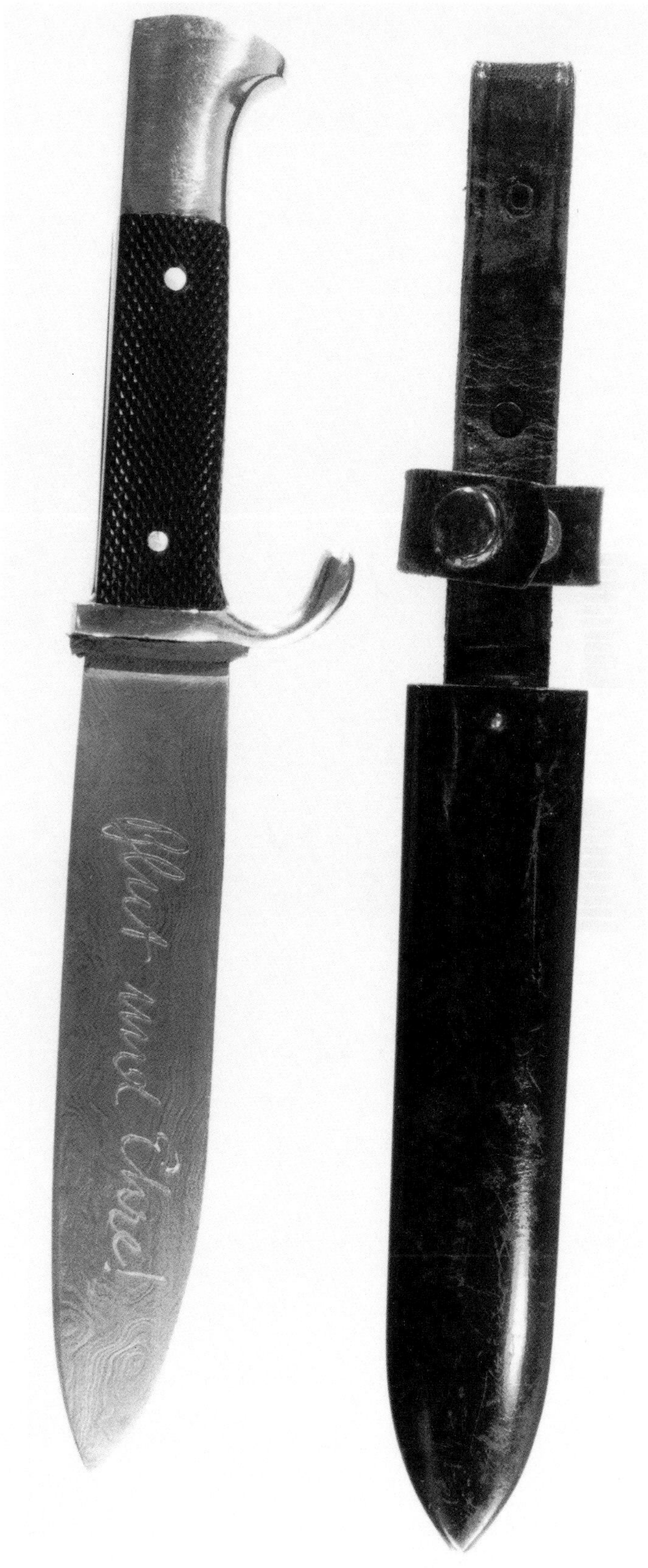

The one-of-a-kind Burg Castle Hitler Youth (HJ) knife is pictured out of the scabbard – reverse full-length view. *Author's collection. Photo by the author.*

boy or his parents could afford to spend in these times of widespread economic depression.[4] Obviously, the original owner of the knife being featured was very well heeled.

All measurements and parts on the Burg Castle HJ knife are standard production with the sole exception of the spectacular Damascus steel blade. What sets the Burg Castle HJ knife apart from the many thousands of other HJ knives that survived the war is the unique, beautifully handcrafted Damast blade with gilt motif. The obverse blade features the huge Burg castle in raised gold near the ricasso followed by a blue ribbon style panel commemorating the 17 September 1933 800th Year anniversary of the castle and surrounded by the typical 1930s German motif of gold oak leaves and acorns. The name of the large well-known Solingen firm producing the knife, "E. & F. Hörster, Solingen" is proudly featured in raised gold letters on the obverse ricasso. The standard Hitler Youth Motto "*Blut und Ehre*!" ("Blood and Honor") is incorporated in raised gold lettering on the reverse blade. It should be pointed out that the motto etching was a facsimile of HJ Leader von Schirach's own handwriting.[5]

Since the Hörster Damascus blade smith found it necessary to move the HJ motto to the reverse side of the blade, he chose to position the motto to be read when the knife is held in the left hand and turned over. While the particular blade maker remains to date unknown, he had to have been one of the foremost Solingen smiths based upon the extreme quality workmanship exhibited by the blade. To state that the *HJ Fahrtenmesser* featured in this article represents the "*crème de la crème*" of HJ knives would be a gross understatement!

Several years ago a German friend of the author residing in Solingen shared a large framed black and white photograph taken during the Third Reich era, which plainly showed a huge *Nazi* flag being flown over the Burg Castle.

Of interest to edged weapon collectors and researchers and highly recommended by the author is the excellent *Schloss Burg Bergisch Museum* which features centuries of important artifacts and includes many historical German sidearms. The *Bergisch Museum* is an important regional museum, as it houses many archeological finds and art treasures from the Middle Ages to the present day. Of special significance is the depiction of the castle during the Middle Ages with splendid examples of the armor used for defense and the property of the lords at the time. A tour of both the entire castle and the museum is available for a very nominal fee.

LEFT: An aerial view of *Schloss Burg*. The Wupper River is located out of sight at the base of the mountain in the foreground. *German postcard.* RIGHT: On the grounds of Burg Castle is an impressive monument to Engelbert II (1185-1225), the Archbishop of Cologne, and, therefore, the Duke of *Westphalia* during the 13th Century. Engelbert was known as "the mightiest man north of the Alps" and built Burg Castle into an insurmountable mighty fortification. *Photo by SGM (Ret.) Kenneth H. Rouse.*

Close-up view of the unique Damascus obverse presentation blade on the E. & F. Hörster Burg Castle Hitler Youth (HJ) knife. The detail of the gilt castle is sufficient to make out both the keep and the battery tower of the castle. *Author's collection. Photo by the author.*

Extreme close-up view of the Burg Castle HJ knife obverse blade celebrating the 800th Anniversary of the castle on 17 September 1933. Note the extreme high quality of the Damascus steel. *Author's collection. Photo by the author.*

Close-up view of the unique Damascus reverse presentation blade on the E. & F. Hörster Burg Castle HJ knife. The standard Hitler Youth motto, *"Blut und Ehre!"* (Blood and Honor) is featured in a raised gold facsimile of HJ Leader von Schirach's own handwriting. *Author's collection. Photo by the author.*

The Ilse Haage *SA* Presentation Dagger

When one thinks of German Third Reich daggers, one's thoughts seldom, if ever, drift to the distaff side. This declaration is not directed towards the distaff collecting of these objects as the names of Jill Halcomb Smith, Emilie Caldwell Stewart, and Julie Brown, come to mind. The real "old timers" in the hobby will also certainly remember the name of Karen Kuykendall of Casa Grande, Arizona, who amassed on her own an *excellent* German dagger collection and wrote extensively about her hobby as a single lady in the 1950s and 1960s.[6]

What is almost impossible to picture is the chauvinistic leaders of the Third Reich presenting one of their "masculine" sidearms to a female! Obviously, it was done at least once, as this article will reveal. Study the face of the presenter, Wilhelm Gustloff, in the picture of his framed portrait reproduced in this article. Frankly, it is hard to picture the "sensitive side" of this individual!

Wilhelm Gustloff was the *NSDAP Landesgruppenleiter, Landesgruppe der Schweiz* (*NSDAP* Regional Section Leader, Regional Section of Switzerland). Although a great deal is known about Section Leader Gustloff, the details concerning him will be purposely kept brief, as this is a story about Ilse Haage, the only woman known to be awarded a presentation *Nazi* dagger.

Wilhelm Gustloff was born 30 January 1895 in Schwerin, Germany, and was murdered by a David Frankfurter on 4 February 1936 in Davos, Switzerland. Gustloff received a Third Reich state funeral in Germany, which was attended by Adolf Hitler.

The Ilse Haage presentation dagger was produced by Hammesfahr & Cie located at 35 Donau *Strasse* in Solingen. This firm, founded in 1919 and registered on 2 May 1919, was formerly the firm of Gustav Hammesfahr. During the Third Reich era early *SA* and *NSKK* daggers were manufactured with the Hammesfahr & Cie "swimmer" trademark and the company name in an oval. Hammesfahr & Cie was also listed among the manufacturers of HJ knives in "*Die Klinge*" Magazine. The firm is currently located at the same address, and a second-generation nephew is presently manufacturing *Besteck* (cutlery).

During the month of October 2003, the author met with the director of the well known *Deutsches Klingenmuseum* in Solingen, Dr. Grotkamp-Schepers, and her assistant, *Frau* Göller, seeking detailed information about the obscure manufacturer of the Haage *SA* presentation dagger, the small Solingen Hammesfahr & Cie firm located at 35 Donau *Strasse*. A telephone call to the current owner of the company had indicated that all old company catalogs, papers, etc. belonging to his great uncle had been donated to the *Klingenmuseum*.

The consensus of the *Klingenmuseum* personnel was that the Hammesfahr & Cie catalogs and papers were not located in the museum archives and the blade dedication was most likely "spurious" since "there was no *Nazi* party organization operating in Switzerland which remained neutral throughout the period." This article will definitely prove otherwise.

As for the blade dedication being spurious, I politely explained that I had seen sufficient German blade dedications in my forty-plus years of studying and collecting these objects to know with some certainty that both the dagger *and* the inscription were 100% genuine.

Yes, the sole presentation dress dagger on record being presented to anyone (male or female) by *NSDAP* Regional Leader Gustloff was the one awarded to Ilse Haage and pictured in this article. Just who was this woman and what did she do to deserve such an honor? Why did *SA* Leader Gustloff single her out to be the recipient of a personalized presentation dagger as well as a framed presentation photograph of himself? Was Ilse Haage a hardcore National Socialist living and working in a neutral country?

The answer appears to be affirmative. *Frau* Haage was born in Hannover and attended the Queen-Luise School in Erfurt. After school she remained in Erfurt but did not learn a profession or pursue work. She met her future husband there and after their marriage the couple moved to Switzerland in 1926. Shortly before Hitler's seizure of power in Germany, Mrs. Haage became a member of the German colony in Bern, Switzerland. On 1 January 1933 she requested entry into the *NSDAP* and received member number 3397879. A *NSDAP* membership book was mailed to her on 30 April 1936. During the year 1934 the couple moved into the city of Zurich and remained there. She was the co-founder of the organization named the German Women's Organization Abroad. In November of 1934 she was trusted with the management of the local Zurich group, whose function was officially confirmed in writing by the local *NSDAP* group leader. During November 1936 she relinquished the leadership and occupied herself exclusively in writing articles about the activity of the German Women's Organization Zurich for the German Newspaper printed in Switzerland. She continued this activity until the end of 1941.

Mrs. Haage attended the 21st course of the *Reichs* School of the *NS*-Women Community in Berlin. Together with her husband Robert Heinrich "Heinz" Haage, she attended the *Nazi* Nuremberg party day rallies in 1935 and 1936. Husband Heinz was a pharmacist in Zurich and also a local leader in the *NSDAP*. Each year until 1944 the couple was able to travel freely to Germany for visitation purposes. The couple paid about twenty-two Swiss francs on a monthly basis for their membership in the *NSDAP*.

As a consequence of the murder of Regional Leader Wilhelm Gustloff on 4 February 1936, The National Territory Switzerland of the *NSDAP* and also the Work Community of the German Women Abroad were forbidden.

Both *Frau* Haage and her husband were eventually expelled from Switzerland during 1945. Ilse Haage remained a dedicated *Nazi* who was proud of her activities until the end. During her investigation she stated the following in her own words, "... My activity of the local group Zurich lasted from November 1934 until November 1936. During this period, I wrote various articles for the German newspaper in Switzerland about the organizations. At the same time I made available a thorough report about the activity of the women's community (in Switzerland) since its existence. I wrote articles for the newspaper until the end of 1941. An entry in my membership book of the *NSDAP* is noted 1 Jan. 1934. This is not exactly correct. I know for certain that I became a member of the *NSDAP* around the middle of 1934. According to an entry in my membership book I became, on the 24th of June, a director of

The Ilse Haage SA presentation dress dagger – obverse view. *Author's collection. Photo by the author.*

DEUTSCHES KLINGENMUSEUM

ABOVE: An English translation of the dedication engraved on the reverse blade of the Ilse Haage presentation dagger reads, "Presented to the District Leader of Zurich, Ilse Haage, in warmfelt camaraderie. W. Gustloff." *Author's collection. Photo by the author.*

BELOW: Extreme close-up photograph of the frosted gothic Ilse Haage SA blade presentation, the W. Gustloff facsimile signature, and the Hammesfahr, Solingen (diver) company trademark. *Author's collection. Photo by the author.*

the *NS* (National Socialist) Women's Work Community, local group Zurich. In the period from the 17th until the 29th of August 1936 I was enrolled in Berlin attending a course for the NS Women's Community. Lectures were given by different individuals whose names I do not recall today. In the years 1935-1936 I was with my husband at the Nuremberg party days. With the same opportunity we visited the meetings of Germans living abroad in Erlangen near Nuremberg. Until 1944 I traveled every year to Erfurt in order to visit my mother and sister. These trips were of a private nature." After the investigation, Ilse Haage was deported on 10 August 1945. The basis for expulsion – local group leader of the *NSDAP* and leader of the German colony.

Although not mentioned in the reports above, *Frau* Ilse Haage was also known as a major fundraiser for the *NSDAP*. Yes, there seems to be very valid rationale for this woman to have received what may well be the only *Nazi* dagger ever presented to a female recipient, creating a true, perhaps one-of-a-kind, rarity in the Third Reich dagger hobby. The only thing more unusual than the Ilse Haage *SA* presentation dagger would be for someone to one day discover an *SS* presentation dagger presented to a female. I don't think that will happen!

OPPOSITE
TOP LEFT: Pictured is the *Deutsches Klingenmuseum* (German Blade Museum) located at Klosterhof 4 in Solingen-Gräfräth. The *Klingenmuseum* has answered countless collector inquiries over the years and hosts an extensive historical Solingen blade production library to which the author has donated numerous original publications from his own 43-year Solingen collection. *Photo courtesy of Dr. Barbara Grotkamp-Schepers.* TOP RIGHT: The present *Direcktorin* (Director) of the *Deutsches Klingenmuseum* is the very attractive and personable Dr. Barbara Grotkamp-Schepers who had the courage to extend the museum's displayed collection through 1945 and the period of the Third Reich. *Photo courtesy of Dr. Barbara Grotkamp-Schepers.* BELOW: This outstanding display of Third Reich era edged weaponry is now on public display at the *Klingenmuseum* in Solingen. The majority of the sidearms were donated by local Solingen residents. *Photo by the author.*

In addition to the presentation 1933 SA dress dagger awarded to *Frau* Ilse Haage, Wilhelm Gustloff also presented her with this 15 December 1935 signed and dedicated photograph of himself, complete with a .800 fine silver frame. The photograph features *Herr* Gustloff in Party uniform complete with Sam Browne belt, swastika armband, Party pin on tie, as well as Party pin pattern cufflinks. *Herr* Gustloff is wearing riding breeches and holding his Kepi. Beneath the photograph on the matte is a three-line inscription to *Frau* Haage signed by Gustloff and dated 15.Dezbr. (December) 1935. The matte also bears the stamp of *Landesgruppe* Schweiz (Switzerland). *Author's collection. Photo by the author.*

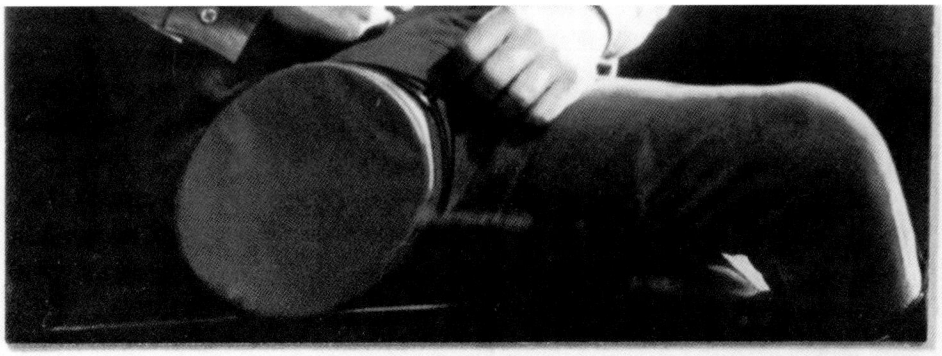

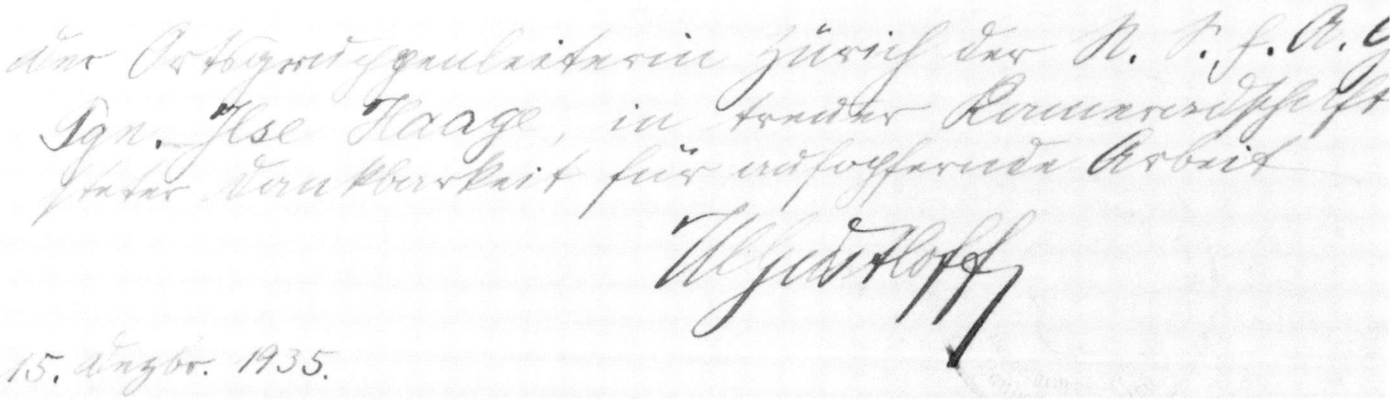

An English translation of the handwritten dedication from Wilhelm Gustloff is as follows: "Dedicated to the female district leader of Zurich, our *NSFAG* party comrade, Ilse Haage in loyal camaraderie and eternal gratitude for your unselfish dedication. W. Gustloff." *Photo by the author.*

N. S. F. A. G. Zürich

Allen deuschen Landsleuten, die in gebefreudiger Weise zur Bereicherung unserer Tombola am „Bunten Abend" zu Gunsten des Winterhilfswerkes beigetragen haben, danke ich im Namen der Frauen-Arbeitsgemeinschaft auf das Herzlichste.

Ilse Haage

Zürich 6, Kinkelstraße 60

An English translation of the above document reads, "I most sincerely thank all German countrymen in the name of the Women's Work Community who contributed in a generous way for the enrichment of our *Tombolo* [Italian word] (lottery game) at the variety show for the goodwill of winter charity." *Johnson Reference Books & Militaria archives.*

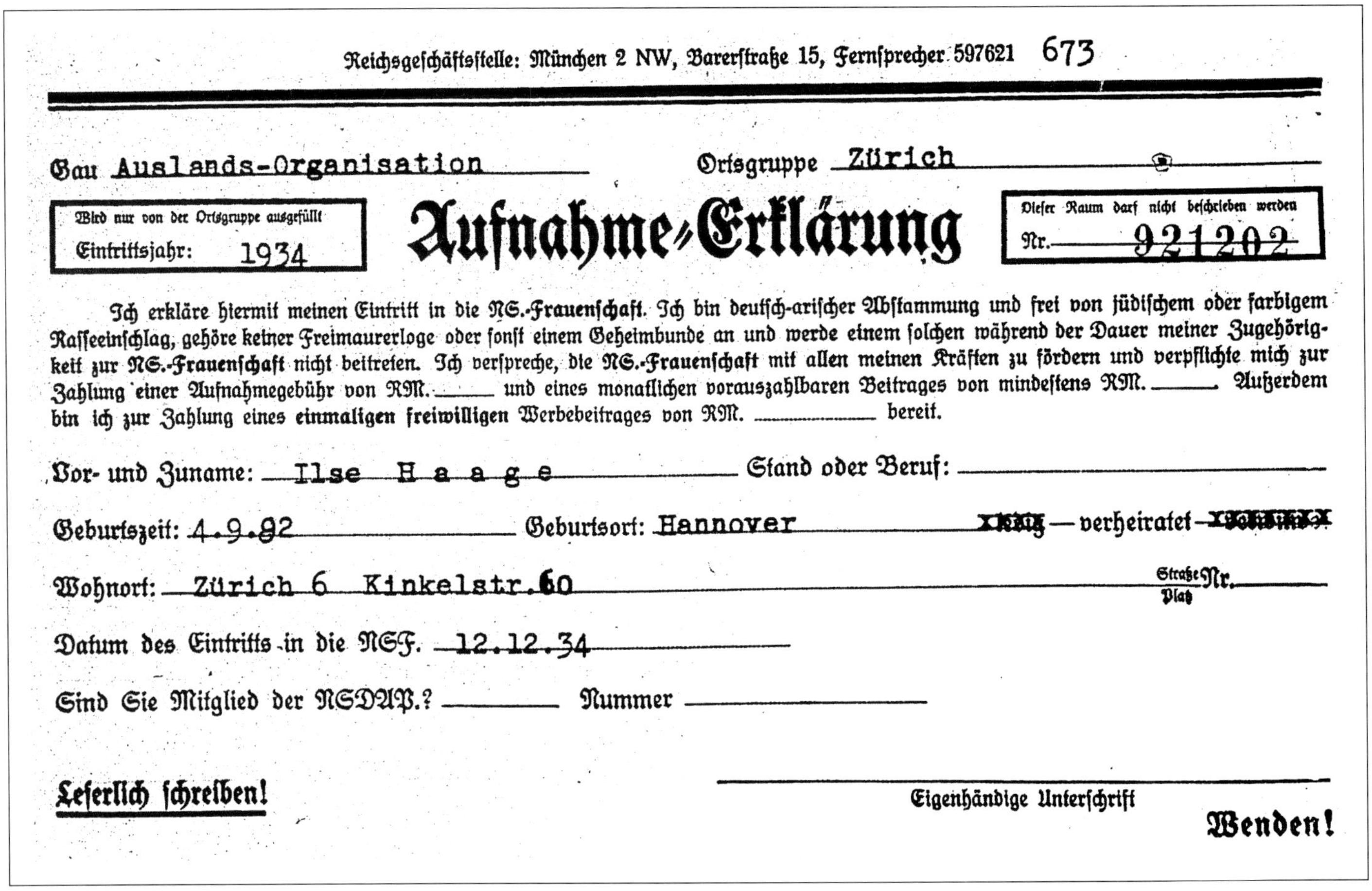

Reichsgeschäftsstelle: München 2 NW, Barerstraße 15, Fernsprecher 597621 673

Gau Auslands-Organisation Ortsgruppe Zürich

Wird nur von der Ortsgruppe ausgefüllt
Eintrittsjahr: 1934

Aufnahme-Erklärung

Dieser Raum darf nicht beschrieben werden
Nr. 921202

Ich erkläre hiermit meinen Eintritt in die NS.-Frauenschaft. Ich bin deutsch-arischer Abstammung und frei von jüdischem oder farbigem Rasseeinschlag, gehöre keiner Freimaurerloge oder sonst einem Geheimbunde an und werde einem solchen während der Dauer meiner Zugehörigkeit zur NS.-Frauenschaft nicht beitreten. Ich verspreche, die NS.-Frauenschaft mit allen meinen Kräften zu fördern und verpflichte mich zur Zahlung einer Aufnahmegebühr von RM. ___ und eines monatlichen vorauszahlbaren Beitrages von mindestens RM. ___. Außerdem bin ich zur Zahlung eines einmaligen freiwilligen Werbebeitrages von RM. ___ bereit.

Vor- und Zuname: Ilse Haage Stand oder Beruf:

Geburtszeit: 4.9.82 Geburtsort: Hannover — verheiratet —

Wohnort: Zürich 6 Kinkelstr. 60 Straße/Platz Nr.

Datum des Eintritts in die NSF. 12.12.34

Sind Sie Mitglied der NSDAP.? ___ Nummer ___

Leserlich schreiben!

Eigenhändige Unterschrift

Wenden!

A copy of the original Enrollment Declaration for Ilse Haage requesting membership in the National Socialist Women's Organization. The entry date on the form was December 12, 1934. In order to become a member, she had to pledge that she was pure German/Aryan descent and was not a member of the Masonic Lodge, etc. *Johnson Reference Books & Militaria archives.*

My sincere thanks to professional librarian Robert McDivitt of Williamstown, New Jersey, who assisted with the research for this chapter, and to Mark Ready of Albuquerque, New Mexico, who helped to translate countless reams of documents from the various Swiss Archives.

The Dr. Wilhelm Nicolaus Damascus *SA* Presentation Dagger

When someone in the edged weapon collecting community exclaims, "they're still out there", this wise adage of advice should not be taken lightly. As recently as this past year's annual Militaria Antiques Xtravaganza (MAX Show) in Pittsburgh, an ultra-rare *SA* Damascus Presentation Dagger with a handmade Damascus steel presentation blade was delivered by a World War II veteran's relative to the author of this article.

I was first made aware of this interesting Third Reich Presentation Dagger several *years* ago when the son-in-law of the previous owner, Mr. Alvin E. Beasley, contacted me via a letter with several enclosed color photocopies of the dagger. The son-in-law was aware of my extensive writing on the subject of German daggers and was, initially, just seeking information on exactly what type of German dagger was in the possession of his family. After an entire plethora of exchanges of correspondence and numerous requests to purchase during the ensuing years, the first hint was dropped late last year by the Beasley family that the dagger possibly was available for sale.

The son-in-law explained that the presentation dagger had been acquired during World War II by his wife's uncle, a U.S. Army corporal who was assigned as the Construction Foreman attached to Company C, 15th Engineer Combat Battalion. Below is an extract from a 19 March 2002 email to the author:

> *Dear Mr. Johnson:*
> Thank you for your interest in the dagger and your continued correspondence. This is what I know about it so far:
>
> My wife's uncle, Corporal John Kruger, landed in France about 6 weeks after the invasion of Normandy. He fought in Europe until the fall of Berlin. He was very interested in German artifacts and technology … to the point (he claimed) of disassembling an ME-262 and putting it into crates, but due to luggage restrictions had to leave it behind. He also saw significant action during the Korean war.

He died about a year-and-a-half ago. My father-in-law was his final caretaker, taking him to chemotherapy treatments, doing his laundry, etc. Much of his estate was left to my father-in-law. My father-in-law was the executor of the estate, and doing inventory of the belongings, found this dagger.

My wife's uncle did accumulate several German artifacts which have attracted the attention of various collectors. One person in particular was interested in the dagger, and initially offered about $400.00 for it. The value of his offers have continued to increase and currently stand at several thousand dollars.

I did not know the translation of any of the German. I don't know the significance of the RZM mark, or the numbers beneath it. (I can e-mail you a better photo of the Eickhorn logo and the RZM mark in the same frame). I know it was presented to a Dr. Nicolaus, but don't know the circumstances of its presentation … or even what events caused such a great knife to be presented. I was hoping you could tell me more of the history of this dagger.

My wife's uncle was always telling war stories. He picked this dagger up somewhere in Europe. I'll ask the family if he ever talked about just how he got it.

/s/

Cpl. Kruger remained enamored with the military after World War II and re-enlisted in the Army on 16 March 1950. He later served in the famous 101st Airborne Division with the rank of sergeant and was discharged from the Army on 7 September 1952.

I found myself in the precarious position of constantly having to better serious offers for the dagger being made elsewhere. At long last, the "magic figure" was finally reached, and the family agreed to sell the Nicolaus *SA* Dagger to me. A deal was sealed with Mr. Beasley, and since his home was in eastern Ohio, he kindly agreed to hand deliver his prize to me at the 2003 MAX Show in Pittsburgh.

The vast majority of the Dr. Nicolaus dagger is "textbook". However, there are two significant differences between this dagger and the other *SA* daggers with Damascus presentation blades on record. One anomaly of the Dr. Nicolaus dagger is the presence of *both* the manufacturer's (Carl Eickhorn) trademark *and* the Eickhorn *RZM* number (M7/66). All of the other *SA* daggers with Damascus blades known to this author feature only one of two of the well-known Eickhorn "squirrel" company trademarks. The likely explanation of this double trademark/*RZM* anomaly is most likely the late date of the presentation of the Nicolaus dagger, 16 May 1937. The *Nazi* Party Department of Ordnance or *Reichszeugmeisterei (RZM)* was established in 1934 in order to exercise strict controls over the manufacture, pricing, and quality control of Party-related items. Any firm that desired to manufacture and sell uniforms, accouterments, and related items was required to acquire official *RZM* sanction and possess an *RZM* manufacturer's control number. After the establishment of the Department of Ordnance, the edged weapon manufacturing firms were assigned unique *RZM* numbers by the *NSDAP* and required to substitute their respective *RZM* number for their company trademark.

Secondly, and equally unique is the fact that the Eickhorn trademark (and *RZM* number) appear on the obverse side of the blade. These company identifications were, obviously, moved to the obverse blade due to insufficient space on the reverse blade due to the lengthy double-line presentation on the reverse blade.

One pertinent question that has not been addressed is, "just who was Dr. Wilhelm Nicolaus?" The credit for running the answer to this question to ground goes to professional librarian Robert McDivitt of Williamstown, New Jersey, for his meticulous research and to Mark Ready from Albuquerque, New Mexico, for his painstaking German/English translations of the countless documents contained in the Wilhelm Nicolaus file.

As it turns out, Dr. Nicolaus was neither a die-hard *Nazi* nor a decorated combatant during World War II. A translation of his condensed service record reflects the following service by Nicolaus during the era of the Third Reich:

SA from 1 January 1933-1945.
SA Sergeant from 4 January 1934-1939.
SA Reserve beginning July 1934.
NSFK from 1936-1939.
R.Kr.B. (Veteran's Organization) from 1935-1945.

A short biographical sketch of the life of Dr. Nicolaus can, perhaps, best be told in his own words. Reproduced below is a careful translation summary of a "Statement of Claim" penned by the doctor on 21 October 1946 in his hometown of Ronsberg, Germany. Since this document is an official, sworn affidavit, accuracy should be generally, if not, completely assumed.

Statement of Claim

Dr. Nicolaus, Wilhelm Manager
Foundation:
Dr. Nicolaus, Wilhelm from 1933-1945
***SA* from 1/1/33 – 1945**
***SA* Sergeant from 4/1/34 – 1939**
***SA* Reserve beginning July 1934**
***NSFK* from 1936 – 1939**
R.Kr.B*(Veterans Association) from 1935 – 1945*
In accordance with the inquiry the "concerned" (Dr. Nicolaus) was brought up in the "Stahlhelm" and later transferred to the *SA*. By the *SA* he was supposed to be brought also to Dachau.

The "concerned" reports:

"In the year 1933 upon advice I entered the party as a member jointly with five other community figures in order to maintain eventually peace (order) in the community and allow no radical movements to rise … I remained in the *SA* …

I was born on the 16th of May 1887 as the second son of the deceased paper manufacturer Heinrich Nicolaus and his wife Emma. I received a good Christian education, attended

the elementary school in Ronsberg, secondary school in Kempten and Weilheim, the industrial school in Augsburg, the technical Colleges in Darmstadt and Munich and the University of Kiel. After my graduation as an engineer I served, 1909/10, with the 7th Field Artillery Regiment in Munich, attained a PhD (Dr. of Engineering), worked as a chemist for the brothers Bayer in Augsburg, and served 1914-18 from the first until the last day in the Army.

In 1917 I married and worked, beginning in 1919, as a partner in the vellum (paper) factory of my father.

When my father died (1917), I became manager of the business ...

From August 1939 until January 1945 I was a participant in World War II.

National Socialism didn't have any inner effect upon my outward life and Adolf Hitler's fanfare brought the first discard to the peaceful work setting and resulted in the division of the work place and community. In order to avoid this, I decided to join the Party. Since I was being denounced having discharged *SA* members from my firm, so in reality, I was at the time already in ill repute in leading National Socialist circles. On June 14, 1933 an "emissary of the Führer" came to arrest me and after excited discussions, he backed off.

This case taught me that it would be also necessary to control the *SA*. So with my "Stahlhelm" comrades I decided to join the *SA* and kept the rank which I had in the "Stahlhelm".

(Later) upon requests of a good acquaintance I joined the *NSFK*. I saw the formation, which concerned itself exclusively with the inexpensive glider sport as a possibility to keep young people away from politics.

I observed as before my religious duties unimpeded went regularly and explicitly to church and to Holy Communion and cared for all the suffering in accordance with Christian ethics and not in accordance with National Socialist propaganda purposes."

/s/
Wilhelm Nicolaus

A second official sworn affidavit dated 3 February 1947 from Markt-Oberdorf states the following:

Beglaubigte Abschrift Markt-Oberdorf, 3.2.1947 Spruch:

Wilhelm Nicolaus was a member of the *NSDAP*, from 1933-1945

of the *SA* Reserve	**from 11/1/33 until 1945**
Sergeant	**from 4/1/34 until 1939**
of the *NSFK*	**from 1936 until 1939**
Veterans Association of the Reich	**1935 until 1945**
of the German Academy	**from 1935 until 1938**

The "concerned" Dr. Nicolaus joined, in the year 1933, the *NSDAP*. At the same time with several other community figures from Ronsberg, since he wanted to impede by his entry, just as the other joining community members, that radical measures would be accomplished by the influence of the National Socialists, and the local official had interest to carry on the community politic further in the former sense. During the time of his party membership, he had not changed views, in any way which rejected National Socialism, but always made an effort to carry out a continual policy in community affairs.

During the time of leadership of the *SA* Reserve troop until his conscription in the Armed Forces in the year 1939 Dr. Wilhelm Nicolaus ensured that the *SA* conducted itself correctly and decently and anti-religious announcements as well as directed acts against the Jews were prevented.

The many witnesses, say that Dr. Nicolaus was never involved with the Party; on the contrary, he never made a secret of his anti-National Socialist attitude.

What does remain totally unexplained is just how Dr. Wilhelm Nicolaus was selected for the presentation of one of the relatively few Third Reich Damascus Presentation daggers bestowed during the period. From the above information, it is obvious that Dr. Nicolaus held neither the rank nor the position in the Party to merit such an award. Additionally, he could hardly be considered a fervent *Nazi* and seemed to operate throughout the entire period on the fringe of the *NSDAP* as a total "non-believer". Therefore, some additional research is needed on this particular treasure before we truly know "the rest of the story"!

Notes

[1] As German edged weapons collectors know, the over-the shoulder Carl Eickhorn trademark is associated with the "Late 1941 to Discontinuation of Dagger production in 1944" (see *Daggers, Bayonets & Fighting Knives of Hitler's Germany*), Major John R. Angolia, page 314. The Grün dagger was presented on 7 November 1941 and is consistent with the Eickhorn trademark dating.

[2] Atwood, James P. *The Daggers and Edged Weapons of Hitler's Germany*. Berlin, West Germany: Omnium Druck and Verlag, 1965, page 61.

[3] Atwood, James P. *The Daggers and Edged Weapons of Hitler's Germany*. Berlin, West Germany: Omnium Druck and Verlag, 1965, page 62.

[4] Lumsden, Robin. *Edged Weapons of Hitler's Germany*. United Kingdom: Airlife Publishing Ltd., 2001, page 49.

[5] Angolia, J. R. *The HJ*. San Jose, California: R. James Bender Publishing, 1991, page 134.

[6] The writer owns to this date a couple of rare Third Reich dress daggers from the Karen Kuykendall collection which were purchased in the early 1960s.

[7] Every attempt, humanly possible, has been made to preclude a single reproduction piece from appearing in this reference, except for photographs appearing in this chapter on reproductions.

[8] The huge Carl Eickhorn weapons factory ceased business in August 1976.

[9] A personal visit by the author to the Höller firm in 1967 verified the manufacturing of Customs dagger reproductions by this firm. The Höller firm has since gone out of business (December 1974).

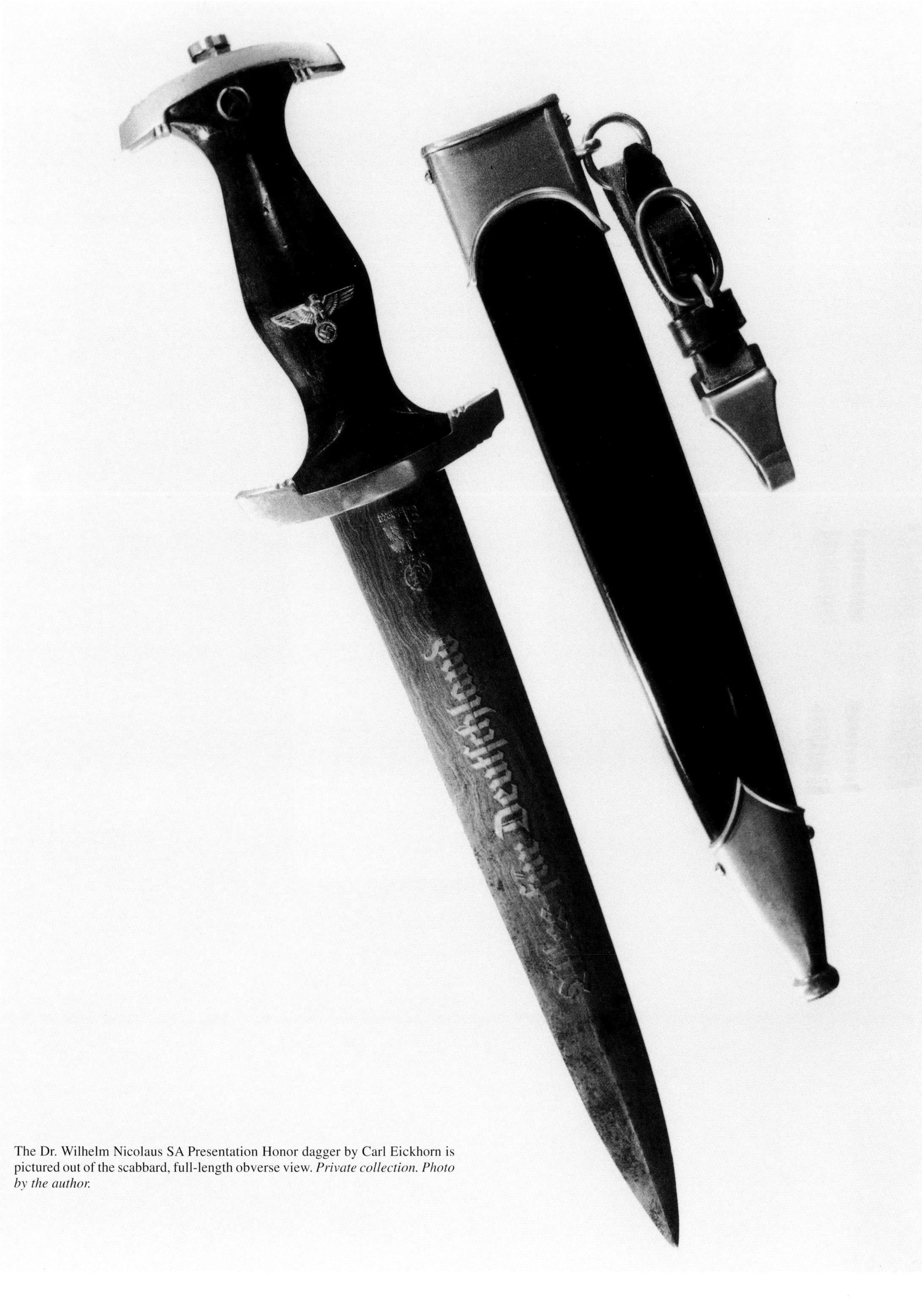

The Dr. Wilhelm Nicolaus SA Presentation Honor dagger by Carl Eickhorn is pictured out of the scabbard, full-length obverse view. *Private collection. Photo by the author.*

Pictured is the raised gold two-line presentation appearing on the reverse Damascus blade of the Dr. Nicolaus SA Presentation Honor dagger. The English translation of the blade dedication reads, "To our troop leader Dr. Wilhelm Nicolaus in fidelity 16 May 1937, Assault Group [or squad] Ronsberg 16/20." Note the high quality "maiden hair" pattern genuine Damascus steel blade. *Private collection. Photo by the author.*

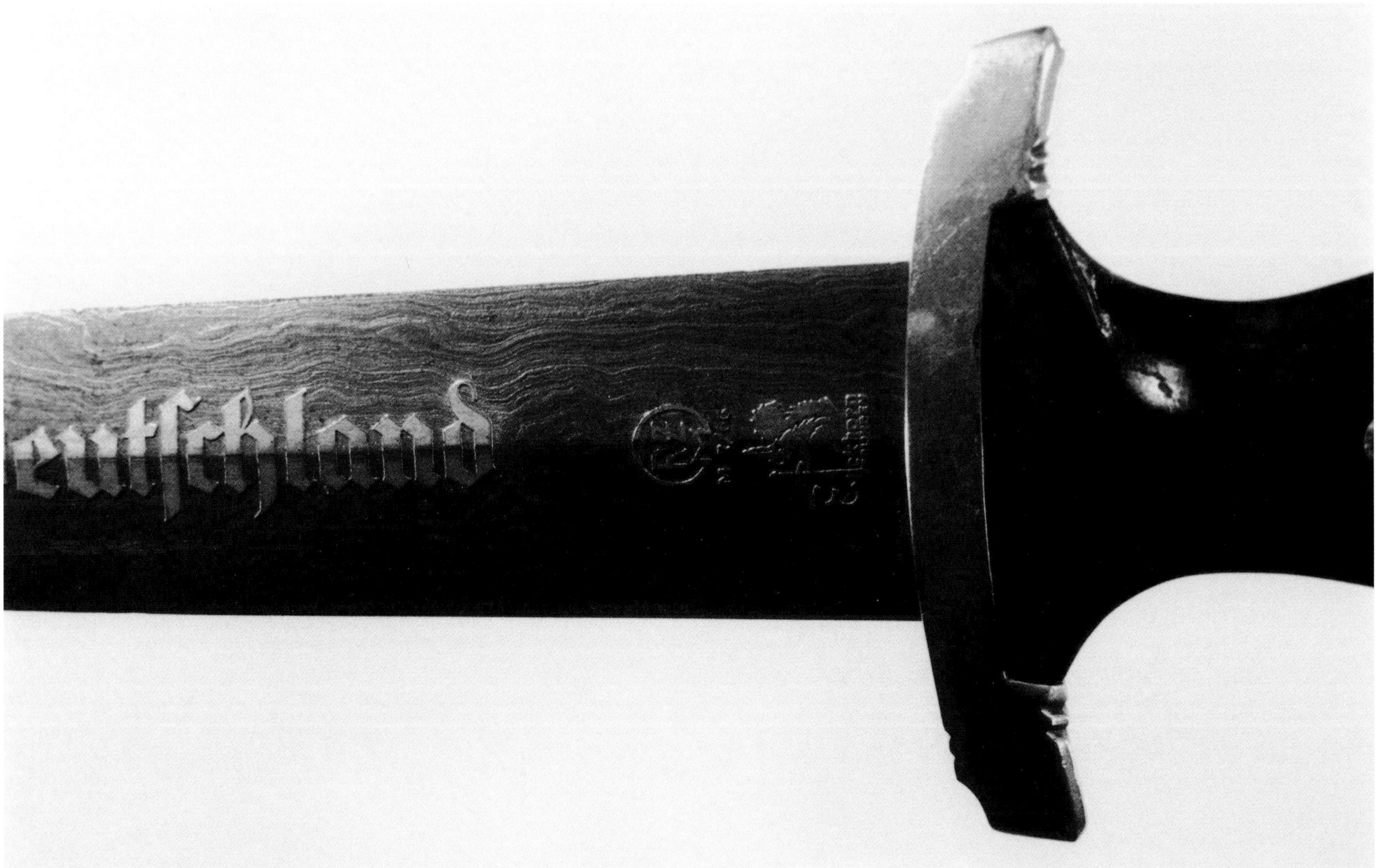

ABOVE: Close-up view of the obverse blade ricasso on the Dr. Nicolaus SA Presentation Honor dagger by Carl Eickhorn. Note the raised gold double Eickhorn "squirrel holding sword" trademark and *RZM* number. These identification marks were obviously moved to the obverse blade due to insufficient space on the reverse blade due to the lengthy two-line presentation. *Private collection. Photo by the author.*

LEFT: Popular dealer/collector/author Ronald Weinand of Quincy, Illinois, and MAX Life Member Carter Jennette of Elizabeth City, North Carolina, examine an ultra rare SA Damascus Presentation dagger that was delivered to MAX co-promoter Tom Johnson by the World War II veteran's best friend at the Radisson Hotel before the 2003 MAX Show. Note the two-line raised gold dedication on the obverse blade. *Photo by the author.*

CHAPTER TWENTY-FOUR

Hermann Göring's Two Bejeweled Third Reich Dress Daggers

"It is an interesting question how far men retain their relative rank if they were divested of their clothes." — Henry David Thoreau

On the subject of dress, Hermann Göring probably had few equals. His peacock flair was well annotated in 1940 by a visiting U.S. reporter who had been invited to *Karinhall* for lunch:

> *"He was dressed in clothes of astonishing originality – long olive-green buckskin boots reaching well above his knees, beige-colored buckskin breeches, a white silk shirt with an unusually large collar and wide, flowing sleeves which tapered to the wrist, and a dark-green sleeveless suede-leather jacket adorned with buttons made from deer's teeth mounted in silver. At the moment we came in, he was proudly calling attention to these buttons, saying that he had shot the deer himself. Around his neck was a red tartan cravat, which was kept in place by a great gold pin, and a gold hunting knife in a gold sheath inlaid with jewels hung from his belt, which had still more gold on it and resembled in many ways, the less costly belts of American cowboys. On the left side of his broad chest he wore the badge of the German Hunting Association, a stag with a swastika resting between its antlers. The stag, too, was gold and the swastika was set with emeralds and diamonds. There were four big rings on his stubby but carefully manicured fingers; the largest, a platinum one set with a ruby and an emerald, was so wide that he could scarcely bend the finger at the second joint."*

Upon his departure from *Karinhall*, the reporter and other guests were presented hunting daggers with the Göring crest, an armored-covered fist clutching a ring, engraved on the hilts.

At no time in history, save the Middle Ages, were personal edged weapons in more evidence than during the Third Reich, and of all the members of the *Nazi* hierarchy, Hermann Göring was by far the most ardent enthusiast of quality blades. In fact, try to locate a photograph of the Number 2 *Nazi* which pictures him in a uniform without a ubiquitous edged weapon of some type at his side. In keeping with his fetish for the absolute *crème de la crème* of material items, the *Reichsmarschall's* personal sidearms were the very finest that skilled Solingen craftsmen were capable of producing.

Reputedly, Hermann's favorite sidearm was a unique hunting dagger (see accompanying photograph) presented to him by his brother-in-law, Count Eric von Rosen. Count von Rosen obviously recognized Göring's gargantuan lust for quality edged weapons, as the Count had the one-of-a-kind stiletto made to order. The hilt and scabbard fittings are skillfully constructed of bronze and silver. An inscription engraved on the front of the straight crossguard translates, "A knife from Eric to Hermann." Highlighting this exotic dagger is the brilliant red leather covering for the scabbard. Numerous photographs of the Reich's Master of the Hunt taken at *Karinhall* depict Göring wearing this particular sidearm. As a matter of fact, he was wearing this piece when he was captured and surrendered it to an intelligence officer in the U.S. Seventh Army. The dagger was subsequently donated to the United States Military Academy Museum at West Point, which graciously provided the close-up photograph for this chapter.

A second unique dagger was tailor-made and presented to Hermann by Count Rosen. The crossguard and pommel of this piece were richly encrusted with precious stones, and it is obviously the dagger that Hermann was wearing at *Karinhall* in the earlier description by the American magazine reporter. The scabbard was designed to reflect Göring's great love for hunting and depicts various hunting scenes. Accenting the precious stones in the hilt is a fluted solid ivory white grip. In order to preclude being completely sans of any Third Reich symbolism, small swastikas were incorporated onto the ricasso of the blade and the upper scabbard fitting.

The U.S. Capture of Hermann Göring

"It was the famous Goering car – a Mercedes Benz with six wheels and bulletproof glass two inches thick. The door opened and Reichsmarschall Hermann Göring, one of the most hated men in Europe, right-hand man to Hitler, bomber of Warsaw, London, Coventry and hundreds of other cities, and the most notorious looter in modern times, lifted his big hulk down from the car ... Göring's wife also got out, and in the dark interior of the palatial car I could hear a little girl crying. Göring was dressed in a plain gray uniform, with a tremendous overcoat of the same color which hung open at the front and exposed his fat figure. He was a short man,

not much more than five feet, but he more than made up for this in breadth. His eyes had something of that glazed expression peculiar to those who take dope. He wore the hat of a Reichsmarschall, but did not carry his marshal's baton. From a belt around his fat middle was suspended a small Mauser Pistol. His face was flabby and heavily lined.

General Stack was introduced to Göring by the aide, and the first question he put to Göring was 'Do you speak English?' Göring answered that he understood the language better than he spoke it. Then, through the interpreter, the mighty Nazi began to apologize for not being dressed in a better uniform. He explained that when the American bombers had leveled Berchtesgaden he had lost most of his uniforms and his medals.

The general and I started to laugh at this display of vanity, but Göring's wife, who had been on the verge of crying since she emerged from the car, burst into tears...

It was almost dark, so we wasted no time starting back. As I turned my jeep around, a group of German soldiers who were watching came up to me. A very young boy who spoke a little English asked if we were going to take Göring away. I said 'Yes, Herr Reichsmarschall is kaput,' at which they burst into laughter. German morale at this time was not very high – nor did our exalted prisoner seem to inspire much devotion."

— *The Saturday Evening Post* Magazine, January 5, 1946

What you have just read is an actual account of the capture of the famous *Reichsmarschall* of the Third Reich, Hermann Göring (or Goering) as reported to *The Saturday Evening Post* magazine in 1946 by the individual responsible for the capture, U.S. Army Captain Harold L. Bond of the 36th Infantry Division, U.S. Seventh Army. The locale for this historic event was a small country road outside of Kitzbuhel, Austria, and the date was May 9, 1945. The credit for Göring's capture goes to Captain Bond's platoon of the 36th Cavalry Reconnaissance Troop.

Hermann Wilhelm Göring was born on January 12, 1893, in the Bavarian town of Rosenheim, Germany, as the ninth son of a local colonial official. Hermann demonstrated an early yearn for military life and attended the cadet college in Karlsruhe. Commissioned as a lieutenant in the infantry in 1912, he served his country during World War I on the western front (*Alsace-Lorraine*). After transferring to the embryonic German Air Force, he distinguished himself in combat and rapidly became a daring air ace by personally downing some thirty-six Allied airplanes. For his superb actions, Göring was awarded Germany's highest decoration for valor, the *Pour le Mérite* (Blue Max).

After Germany collapsed, the young aviator fled to Sweden and for two years was employed as a commercial pilot and mechanic for Swedish Air Lines. Returning to his native Germany in 1922, Hermann Göring joined the *Nazi* party and became one of the earliest supporters of an unknown named Adolf Hitler. In 1928, Göring became one of the first *Nazis* to be elected to the *Reichstag*, and in four short years, become its President.

When Hitler became Chancellor on January 30, 1933, and in complete control of Germany, what happened to Hermann Göring's

OPPOSITE: On the subject of dress, Hermann Göring had few equals In this "Life" Magazine color photograph he is dressed as the "Master of the Hunt." Prominently displayed on his left side is his favorite dagger, given to him by his Swedish brother-in-law Count Eric von Rosen. Incongruously, Göring wore this same dagger with his military uniform when he surrendered to American forces in 1945. Its present day location is the West Point Military Academy Museum in New York State. *"Life" Magazine photograph. Photo courtesy of N. C. "Dutch" Heilman.*

ABOVE: Previously unpublished photograph of the Mercedes-Benz automobile that delivered Hermann Göring to the members of U.S. Army Captain Harold Bond's platoon outside of Kitzbuhel, Austria, on 9 May 1945. *Keith Wilson, Jr. collection. Photo by Floyd A. Jacob.*

career is now history. He was immediately named as a minister to the *Führer's* cabinet. As Prussian Premier and Minister of the Interior, Göring was tasked with organizing the *Nazi* Secret Police (*Gestapo*) and creating the first concentration camps. Recognizing a weakness in his air arm, Hitler charged Göring with completely overhauling the floundering German Air Force. And overhaul he did! As *Reichsminister* of Aviation, Hermann Göring organized and expanded the *Nazi Luftwaffe* into one of the most powerful and modern air powers in the world in only two short years. From 1933 until the *Nazi* collapse in 1945, he remained in supreme command of the *Luftwaffe*. During this same 1933-1934 timeframe, the *Reichsminister* began his industrial ventures in Germany and abroad and reputedly became one of the richest men in the world.

In 1940, Göring was promoted to the unique rank of *Reichsmarschall* of the Greater German Reich and designated a member of the Cabinet Council. The following year he became chairman of the secret Defense Council and was made president of the Council of Economic Warfare.

Hermann Göring has been called many things by many people. Opinions of Göring in the world press during the Third Reich regime ranged from "a brilliant, debonair, sophisticated male" to "an overweight, egotistical swine." However, one constant comes to the fore in all biographical studies of this controversial individual – the man loved luxury and what money could buy. Consider for a moment Göring's lavish, 100,000-acre hunting retreat "*Karinhall*" in the *Schorfheide* Forest some thirty miles from Berlin. This beautifully situated retreat, named for Hermann's first wife, was complete with a gymnasium, rifle range, and bowling alley. In jest, the *Reichsmarschall* referred to the sumptuous manor as "my little weekend house." Göring elected to travel the thirty miles to Berlin in a racy, sky-blue, sixteen-cylinder compression sports car, e.g., the image demanded a sports car even in the 1930s!

Göring's rise to power is a study in speed. During this short emanation, he held a plethora of significant titles including Commander of the "Flying Circus," President of the *Reichstag*, the Reich's Master of the Hunt, *Reichsminister* for Prussia, Supreme Commander of the Air Force of the Reich, Air Minister, *Reichsmarschall* of the Greater German Reich, and President of the Council of Economic Warfare. In late 1939, the *Führer* even named Göring as his potential successor. The rapidity of Hermann's rise to the position of the Number 2 *Nazi* was exceeded only by his decline. After 1943, Göring's reputation declined after each Allied

Reichsmarschall Hermann Göring arrives on 9 May 1945 for detention at the camp in Augsburg, Germany, where he is interviewed by Colonel William W. Quinn, U.S. Seventh Army G2 officer and Major Paul Kubala, Assistant G2 officer, U.S. Seventh Army interpreter. *U.S. Army photo.*

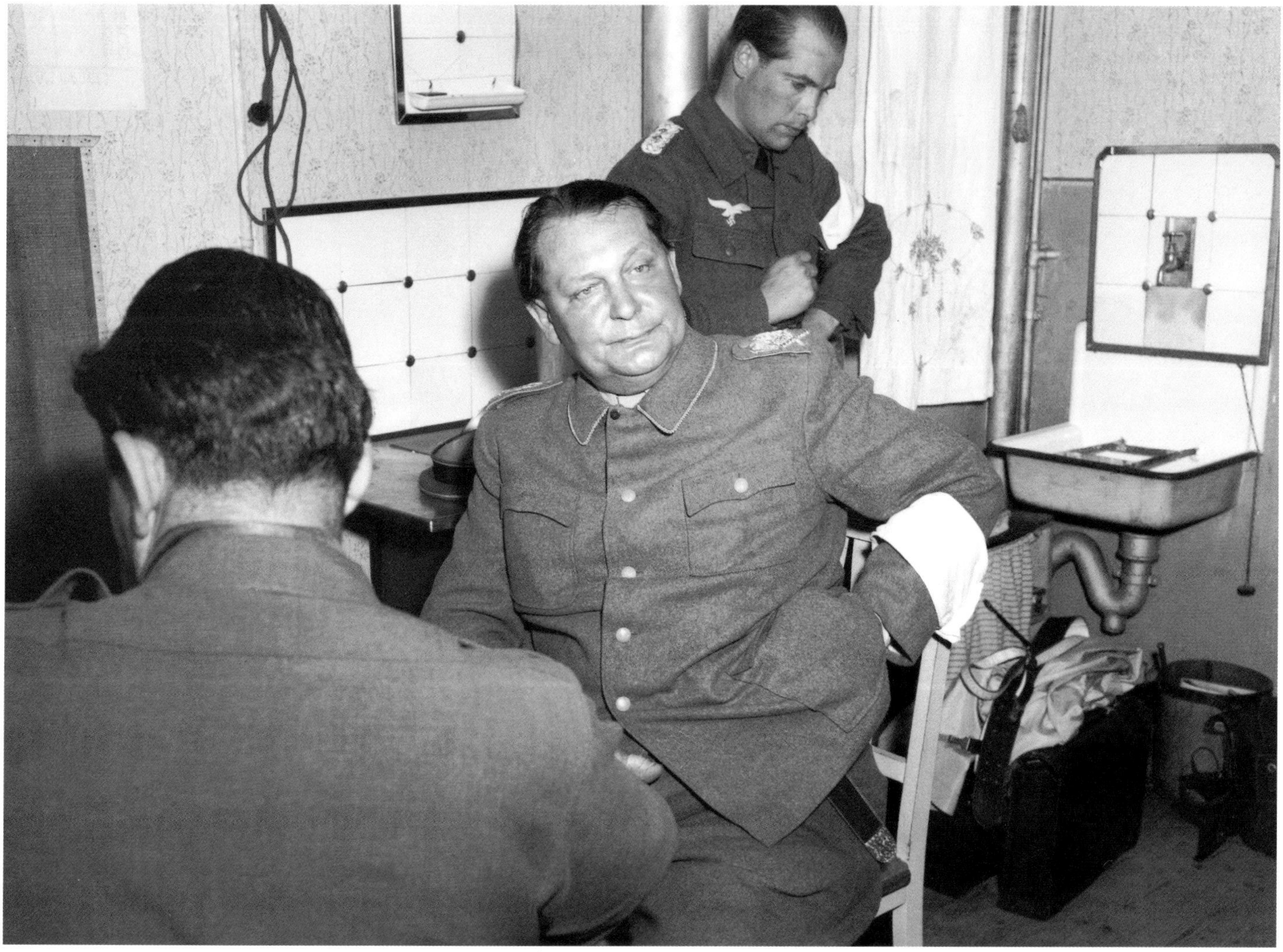

Reichsmarschall Hermann Göring is pictured removing his favorite hunting dagger while being interrogated by G2 officers at the Augsburg, Germany POW camp. Göring's aide-de-camp, Colonel Berndt von Brauchitsch is seen in the background. The hunting dagger is now on public display at the U.S. Military Academy at West Point, New York. *U.S. Army photo.*

air raid, and his close relationship with Hitler quickly deteriorated to a state of atrophy. During late April 1945, the heir-designate of the *Führer* resigned and fled to the Alpine mountain retreat of Berchtesgaden where he had previously stashed out a collection of art, which was valued in excess of $200 million! His Herculean economic duties were turned over to Dr. Walter Funk, Minister of Economics, and Professor Albert Speer, Minister for Arms and Munitions.

After his capture on May 9, 1945, by the U.S. Army, Göring was sent with his personal aide-de-camp Colonel Berndt von Brauchitsch to Augsburg, Germany, to be detained in a prisoner-of-war camp while awaiting trial before the International War Crimes Tribunal at Nurnberg on November 20. Although he entered a plea of "Not Guilty" at his trial, Göring was found guilty on October 1, 1946, to all four charges (e.g., conspiracy to wage aggressive war; crimes against peace; crimes violating laws of war; and crimes against humanity) and sentenced to be hanged. A hidden vial of poison precluded Göring's trip to the gallows, and his body was cremated on October 16, 1946, at the former Dachau Concentration Camp just north of Munich, Germany.

The Hermann Göring Bejeweled Hunting Dagger

Author's Note: Undoubtedly, the 'Crème de la Crème' of Third Reich dress daggers are the two bejeweled daggers commissioned for and belonging to Luftwaffe ace Hermann Göring. Both daggers are legendary in the annals of dagger collecting and have been covered in detail in the ongoing eight-volume series, "Collecting the Edged Weapons of the Third Reich" by the author. When contemporary edged weapon collectors think of these two fabulous one-of-a-kind sidearms, not one, but two names come to mind – Hermann Göring and the late Ben Swearingen – Göring for owning the daggers and the late Ben Swearingen for locating not one, but both pieces! Ben was easily the most successful of all of the Third Reich treasure hunters having the Göring jeweled daggers, the Sepp Dietrich SS Honor sword, numerous Adolf Hitler personality items to his credit, as well as countless other Third Reich artifacts that he personally traced down. Ben served as a major contributor to Volume V and Volume VIII of "Collecting the Edged Weapons of the Third Reich" and Volume II of "World War II German War Booty". Unfortunately, Ben succumbed to a long bout with a brain tumor during 1998. The entire militaria hobby will sorely miss this exceptional individual with a collector's zeal and a detective's instincts to ferret out the most valuable German war booty to survive World War II. What are the sheer odds that the same collector/ researcher would be involved in the discovery of both of the two most sought-after German dress daggers in existence? The remainder of this chapter will be devoted to repeating, in Ben Swearingen's own words, the two stories for the record, which are published as submitted with little editorial intervention.

Ben Swearingen

With regard to the Göring Hunting dagger, I wish that I was able to relate for your readers the details of a thrilling hunt and chase. But the truth is that there was no such thing, as the dagger almost fell into our hands. For what it is worth, I'll give you the details of the find.

In 1975, rumors of the whereabouts of the dagger reached Ben Curtis and myself through some comments made by a third person at the Houston Gun Show. We learned that a "jeweled cutlass", which supposedly had belonged to Göring, was in the possession of the widow of a former U.S. Army officer in a small town in Arizona *[Editor's note: Kingman]*. A telephone call to this lady proved that she did, indeed, have some sort of dagger – which she attributed to Göring – as well as other souvenirs. Some time later, she sent us color photographs of the entire collection. We immediately realized from the photographs that we had, indeed, located the fabulous Göring Hunting dagger, lost since the Army inventory at Berchtesgaden in 1945.

Ben and I immediately booked a flight to Las Vegas, where we rented a car and drove to the sleepy Arizona town. After dinner with the lady, we returned to her home and began our negotiations for the dagger and the rest of the collection, none of which we had yet seen. We discussed the matter until well past midnight, and the lady still would not agree to the price we offered. As we were not prepared to offer a higher figure, we dejectedly returned to our motel, convinced that the Göring dagger had eluded us. But the lady had a change of heart, calling us at 5:00 a.m. to say that she would accept the figure we offered. Ben and I then spent some anxious hours waiting for the local bank, where the souvenirs were stored, to open.

At the bank, we descended the steps into the vault. In a few minutes, the lady was bringing forth box after box, all filled with Göring relics. She handed me a box from which I withdrew a homemade red-and-black checkered cloth pouch. Unzipping the top of the pouch, I soon had the Göring Hunting dagger in my hand. It was more beautiful than I expected, but I immediately noticed that the ivory grip was loose.

Setting the dagger aside for a moment, we examined the other items. There was a magnificent thirty-two piece set of the finest Sevres china, a huge silver baton studded with an arabesque design of Persian turquoise, some Hitler and Göring silverware, and other miscellaneous items. One unusual item was a cased Field Marshal's baton from the period of the French Restoration, evidently from Göring's own collection of militaria. All of these items had been "liberated" from the treasure caves at Berchtesgaden, and sent back to the States by the lady's husband. We transacted our purchase, gathered up our booty, and drove back to Las Vegas.

Getting back to Texas with the dagger proved to be more of a problem than we had anticipated. We were told that we could not carry the dagger onto the airplane with us, but had to store it in the baggage compartment. Needless to say, Ben and I were not willing to take the chance that the dagger, after being "lost" since 1945, might again go astray. We solved our problem by showing the dagger to an airline official, who ironed out all problems for us. Not only were we allowed to carry the dagger onto the airplane, but we were escorted by armed guards. The dagger was placed in the cockpit during the flight, and at Dallas we were escorted off the plane by more armed guards. We returned to our homes, dead tired, but happy.

The next problem we faced was convincing anyone that we actually had the dagger. I offered the story and exclusive color pictures of the dagger to a collectors' magazine, only to have the offer turned down flat! Clearly, the editor thought that another spurious dagger had been "discovered". However, a few newspaper articles – one on the front page of the Dallas *Morning News* – and the exchange of some photographs with various collectors changed all that. Suddenly, the dagger had become a "hot item", and various collectors were calling and writing. Most were shocked into silence by what was then considered a very high price.

One person was not scared off by the price – the late Roger Steele of Hollywood Military Hobbies fame. Roger wanted to see the dagger, and I flew to Las Vegas to meet with him. I could tell that Roger was "hooked" the minute he saw the dagger. After lengthy negotiations – Roger spent hours telling me why he couldn't pay the price we were asking – he agreed to our price. Some months later, he paid in full, and the dagger went to him.

It is my understanding that the dagger sold again after Roger's death, and is now in the possession of a collector. I am sure that the thrill of ownership for him is great, but it could not match the thrill that Ben and I experienced in locating the dagger, our "greatest find" to date. I hope this is of some use for you, Tommy. As I said, it is not the greatest "chase" story imaginable, but it is the true story of how the dagger was found.

Editor's Note: The Göring Hunting Cutlass was sold by the late Roger Steele in the late 1970s to advanced collector Dr. Ron Distelhorst of Schaumburg, Illinois. The dagger reposed in the extensive Distelhorst collection until 1992 when it was purchased by the late Robert Waitts of Paterson, New Jersey.

OPPOSITE: Beautiful close-up study of the one-of-a-kind Hermann Göring bejeweled Hunting dagger. This richly decorated hunting dagger, presented to Göring by his brother-in-law, Count Eric von Rosen, features numerous precious stones, to include diamonds. The scabbard was designed and handcrafted to reflect Göring's great love for hunting and depicts various hunting scenes. Accenting the precious stones in the hilt is a fluted solid ivory white grip. Note the incorporation of a small swastika on the blade ricasso and the Göring family coat of arms on the scabbard. Period photographs of Göring wearing this dagger are available on pages 972, 975 and 976 of this reference. *The late Robert Waitts collection. Photo by Charles H. Jenkins, III. Photo courtesy of Thomas T. Wittmann.*

The ornate one-of-a-kind Hermann Göring bejeweled Hunting dagger is pictured – full-length obverse view. Note the small swastika incorporated into the obverse blade ricasso. *The late Robert Waitts collection. Photo by Charles H. Jenkins, III. Photo courtesy of Thomas T. Wittmann.*

The one-of-a-kind Hermann Göring bejeweled Hunting dagger is pictured – full-length reverse view. The dagger was constructed to allow either side to be worn exposed. Note that an additional swastika has been included in the scabbard design. *The late Robert Waitts collection. Photo by Charles H. Jenkins, III. Photo courtesy of Thomas T. Wittmann.*

ABOVE: Beautiful close-up study of the one-of-a-kind Hermann Göring bejeweled Hunting dagger. Although this spectacular sidearm has been covered in detail in the previous works by the author, this is the first time that the reverse scabbard has been shown. This richly decorated Hunting dagger, presented to Göring by his brother-in-law, Count Eric von Rosen, features numerous precious stones, and period photographs of Göring wearing the dagger are available on pages 972, 975 and 976 of this reference. *The late Robert Waitts collection. Photo by Charles H. Jenkins, III. Photo courtesy of Thomas T. Wittmann.*

OPPOSITE: Close-up study of the reverse hilt and upper scabbard of the ornate one-of-a-kind Göring bejeweled Hunting dagger. The scabbard is covered in a high-grade green-pebbled leather with gold hunting scenes superimposed. *The late Robert Waitts collection. Photo by Charles H. Jenkins, III. Photo courtesy of Thomas T. Wittmann.*

Extreme close-up view of the pommel of the one-of-a-kind bejeweled Hermann Göring Hunting dagger. The pommel cap is encrusted with genuine diamonds and rubies. The Göring "mailed fist and ring" coat of arms is deeply etched into the large crystal decorating the top of the pommel. *The late Robert Waitts collection. Photo by Charles H. Jenkins. Photo courtesy of Thomas T. Wittmann.*

One view of the spectacular Hermann Göring bejeweled Hunting dagger which has not been shown in any of my previous reference books is the top of the pommel which features the intricate incorporation of the Göring family *Wappen* (coat of arms), a mailed fist clutching a ring. *The late Robert Waitts collection. Photo by Charles H. Jenkins, III. Photo courtesy of Thomas T. Wittmann.*

AN INTERESTING DESK STAND USED BY GOERING. THE "INKWELL" ON THE LEFT HOUSES A SECRET DICTAPHONE, WHICH COULD TELL MANY STRANGE TALES.

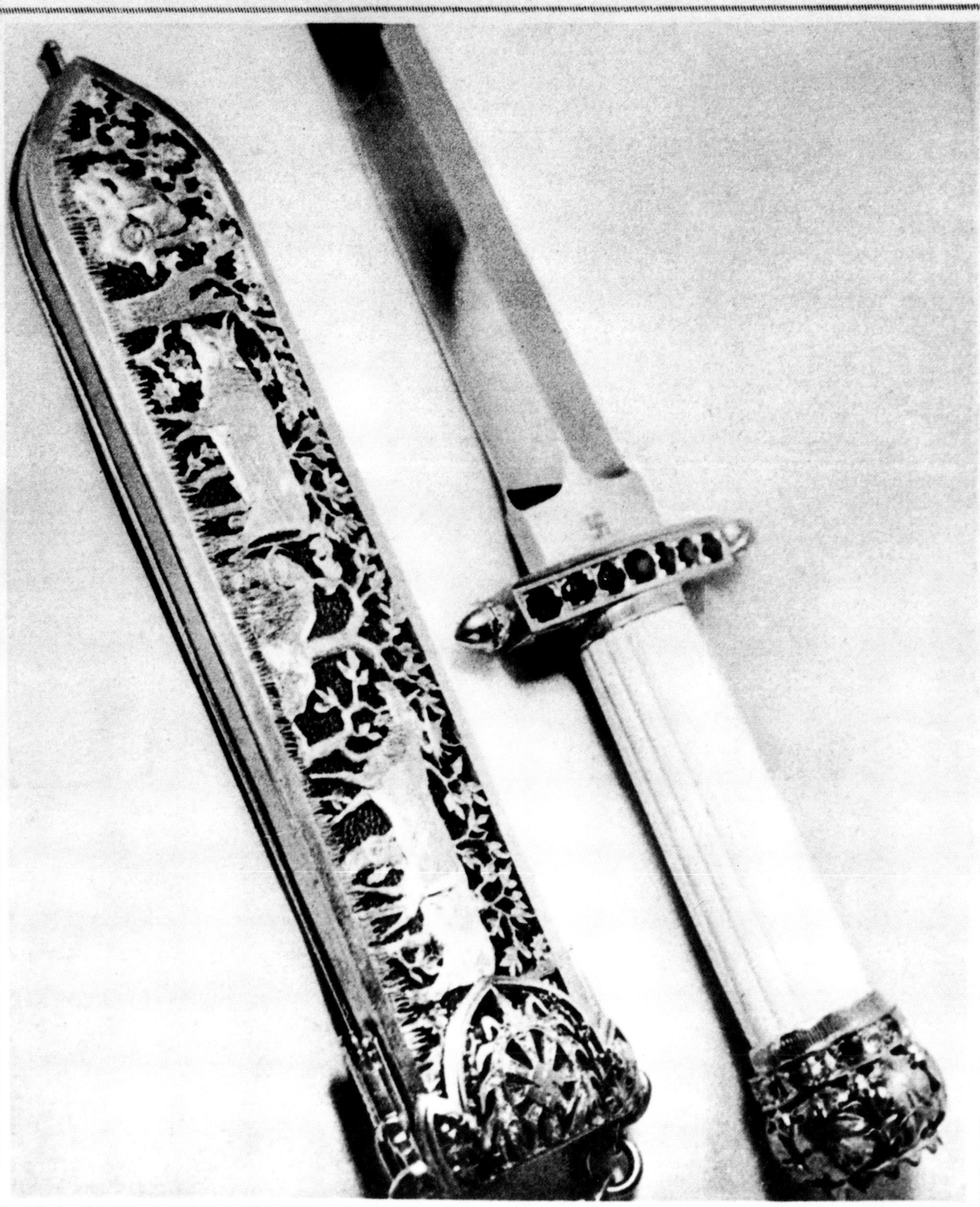

ONE OF THE HUNTING DAGGERS AND SCABBARDS WORN AND USED BY GOERING. ITS MATERIALS INCLUDE SILVER-GILT, TOURMALINES, AND GREEN LEATHER.

A 24 November 1945 photograph of the ornate, one-of-a-kind Hermann Göring Hunting dagger. *Photo courtesy of "The Illustrated London News" picture library.*

584—THE ILLUSTRATED LONDON NEWS—Nov. 24, 1945

MADE IN GERMANY: OSTENTATIOUS TREASURES FROM GOERING'S "ALADDIN'S CAVE"; PRODUCTS OF BERLIN TECHNICAL ACADEMY.

A U.S. SERGEANT EXAMINING ONE OF MANY ORNATE "WORKS OF ART," MADE IN BERLIN TECHNICAL ACADEMY FOR GOERING AND FOUND IN HIS PRIVATE HOARD IN A MOUNTAIN CAVE. ANOTHER PHOTOGRAPH SHOWS THE BOX IN DETAIL.

THIS BOX, NEARLY 14 IN. LONG, MADE BY STUDENTS OF BERLIN TECHNICAL ACADEMY, HAS AN ORNATE KNOB OF SILVER-GILT AND AQUAMARINES.

AN INTERESTING DESK STAND USED BY GOERING. THE "INKWELL" ON THE LEFT HOUSES A SECRET DICTAPHONE, WHICH COULD TELL MANY STRANGE TALES.

DESIGNED AS A CHURCH SHRINE, THIS HEAVILY GILT SILVER CABINET WAS USED BY GOERING AS A COCKTAIL CABINET. IT IS LINED WITH MIRRORS.

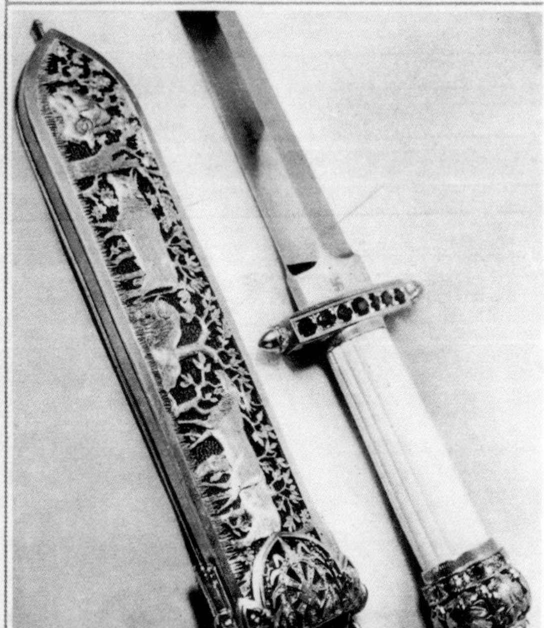

ONE OF THE HUNTING DAGGERS AND SCABBARDS WORN AND USED BY GOERING. ITS MATERIALS INCLUDE SILVER-GILT, TOURMALINES, AND GREEN LEATHER.

Among the many treasures of Goering's hoard, of which examples have already appeared in our pages, there has been found a number of pieces specially designed and produced for him by the Berlin Technical Academy. Made by students of the Higher Class of the Academy, to the design of its director, Professor Herbert Zeitner, who himself put the finishing touches to many of them, they reflect the crude tastes of their former owner. In some cases, indeed, the decorations ordered were so ostentatious that they were constructed to be easily removable, thus leaving a basic design a little closer to a simpler form of art. An example of this is the table-centre of the Lucia-Bride, whose ornate necklace, of silver-gilt mounted with moonstones, is a separate piece. The Lucia-Bride is the central figure of a Swedish custom on the festival of St. Lucia (December 13), when the daughter chosen for this honour, wearing candles in her hair, awakens the

ABOVE AND OPPOSITE: Pictured on two pages of photographs are the many treasures of the Göring hoard from the "Aladdin's Cave" in Berchtesgaden which appeared in the 24 November 1945 issue of "The Illustrated London News" on pages 584 and 585. Included in this grouping specially designed and produced for Göring by the Berlin Technical Academy under the Direction of Professor Herbert Zeitner are the two Göring bejeweled dress daggers. *"The Illustrated London News" picture library. Photo courtesy of Barry Brown, United Kingdom.*

Nov. 24, 1945—THE ILLUSTRATED LONDON NEWS—585

(ABOVE.) THIS GREEN LEATHER BELT, WITH ITS GILT CLASP AND DECORATIONS, ON WHICH ARE MOUNTED PRECIOUS STONES, WAS WORN BY GOERING WITH ONE OF HIS EXTRAVAGANT HUNTING SUITS.

(LEFT.) ANOTHER PRODUCT OF ACADEMY STUDENTS TO GOERING'S ORDERS: A BOX IN WHICH SILVER-GILT, GOLD, PEARLS AND PRECIOUS STONES ABOUND, WITH ORNAMENTS MOUNTED ON IVORY PLAQUES.

A LARGE SILVER-GILT TABLE-CENTRE, DECORATED WITH PRECIOUS STONES. IT REPRESENTS THE LUCIA-BRIDE—THE CENTRAL FIGURE IN SWEDISH CELEBRATIONS OF THE FESTIVAL OF ST. LUCIA.

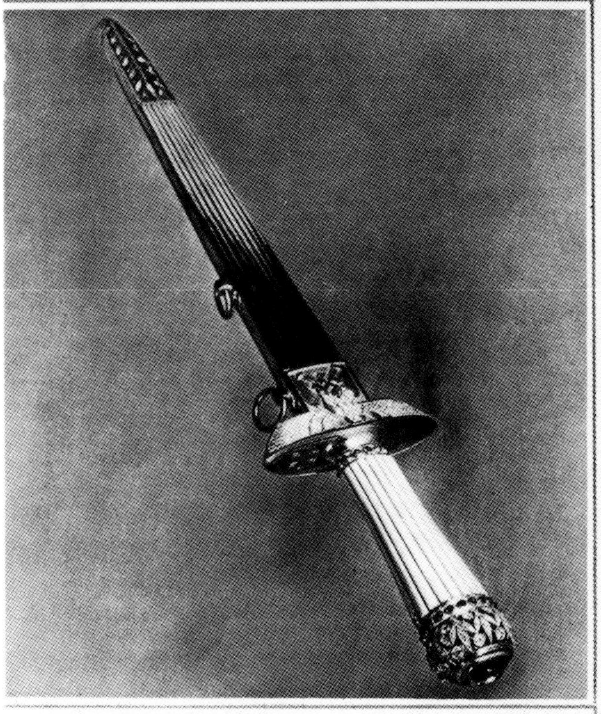

GOERING'S MARSHAL-OF-THE-REICH "DAGGER OF HONOUR" IN ITS SCABBARD. PROFUSELY ORNAMENTED WITH DIAMONDS, IT WAS MADE BY ACADEMY STUDENTS.

A DETAIL VIEW OF THE TABLE-CENTRE SEEN ABOVE. THE NECKLACE IS REMOVABLE, AS ARE THE MORE CRUDE FEATURES OF SOME OTHER PIECES.

household and distributes food. She is addressed only as "Lucia-Bride" on that day, and not by her own name. This figure was one of the designs of Professor Zeitner, who, before becoming director of the Higher Class in the Academy, was director of gold and silver crafts at Berlin High School, and who, while carrying out the orders of his political masters, at the same time strove for basically sound art structure. In several of the pieces illustrated, this "pure" design is visible beneath the trappings. Of the other pieces illustrated, the decorations in general are a reflection of Goering's notoriously vulgar ostentation, and the hunting accessories in particular of his hankering after the barbaric splendours of the Middle Ages. One other aspect of Goering's character is revealed in the innocent-looking desk-stand, whose concealed dictaphone could record the words of his visitors and guests.

This ornate green leather belt, with its gilt buckle and decorations, on which are mounted precious stones, was worn by Göring with one of his many extravagant hunting outfits. *"The Illustrated London News" picture library. Photo courtesy of Barry Brown, United Kingdom.*

The ornate Hunting dagger in wear. Göring appears to be displaying this prize to the photographer. The scene is his hunting estate, *"Karinhall."* The Naval officer on the right wears the standard Navy dirk, complete with portepee. *Library of Congress photo.*

DALLAS TIMES HERALD, Thurs., Dec. 25, 1975 ★★★★ B—13

—Staff Photo

Ben Swearingen displays Nazi weapon

Texans purchase Goering cutlass

By ALICE LOVE
Staff Writer

Two Texas collectors of World War II relicshave paid over $40,000 for a hunting cutlass used by Herman Goering,head of the Nazi air force during Adolf Hitler's Third Reich. The cutlass had been reported missing since 1946.

Ben Swearingen of Denton and Mike Giles of Dalls who have been collecting German war souvenirs since 1948, said the acquisition was 'one of the most significant finds in the past 30 years."

Giles said the cutlass had been photographed and catalogued in collectors' manuals in 1946 and was listed as missing until he and Swearingen purchased it two weeks ago.

Swearingen and Giles said they first learned of the dagger's location through the stepdaughter of an American colonein couner-intelligence who had been stationed n Germany during World War II.

When they contacted the American colonel's widow in uscon, Ariz., she told them that her husband had taken the cutlass and other relics from Goering's summer cottage at Obersalburg, German following the war.

The gold and silver dagger, encrusted with the Goering family crest, is mounted with several dozen precious stones, including rubies, diamonds and spalines, said Giles.

Swearingen and Giles also bought a 32-piece set of Sevres china with the Goering coat of arms on each piece, a seven-piece silver service owned by Goering and 25 pieces of silverware with Hitler's crest from the woman in Arizona, who did not want her name publicized.

⅛Giles would not reveal the price paid for the collection, but he said all the pieces had been authenticated and catalogued and were 'priceless—one of a kind."

Swearingen's and Gile's etensive colletion of Third Reich relic also includes the key to the City f Munich presented to Hitler during the war, a 14K gold fountain pen given by Hitler to his mistress, Eva Braun, and several dozen German swords and daggers used during the war.

LEFT: A black and white study of the Hermann Göring bejeweled Hunting dagger – obverse view. This rare, one-of-a-kind dress dagger has been featured in three different volumes of the *Collecting the Edged Weapons of the Third Reich* series by the author. The dagger is extremely well documented, and was crafted by the Zeitner jewelry firm in Berlin and presented to Hermann Göring by his brother-in-law, Count Eric von Rosen. The crossguard and pommel of this unique piece are encrusted with semi-precious stones (what appears to be alternating zircons and tourmalines). The only marking to be found on the exterior of this dagger is the maker's name, "Zeitner, Berlin" which is stamped on the side of the scabbard. *The late Robert Waitts collection. Photo by the author.*

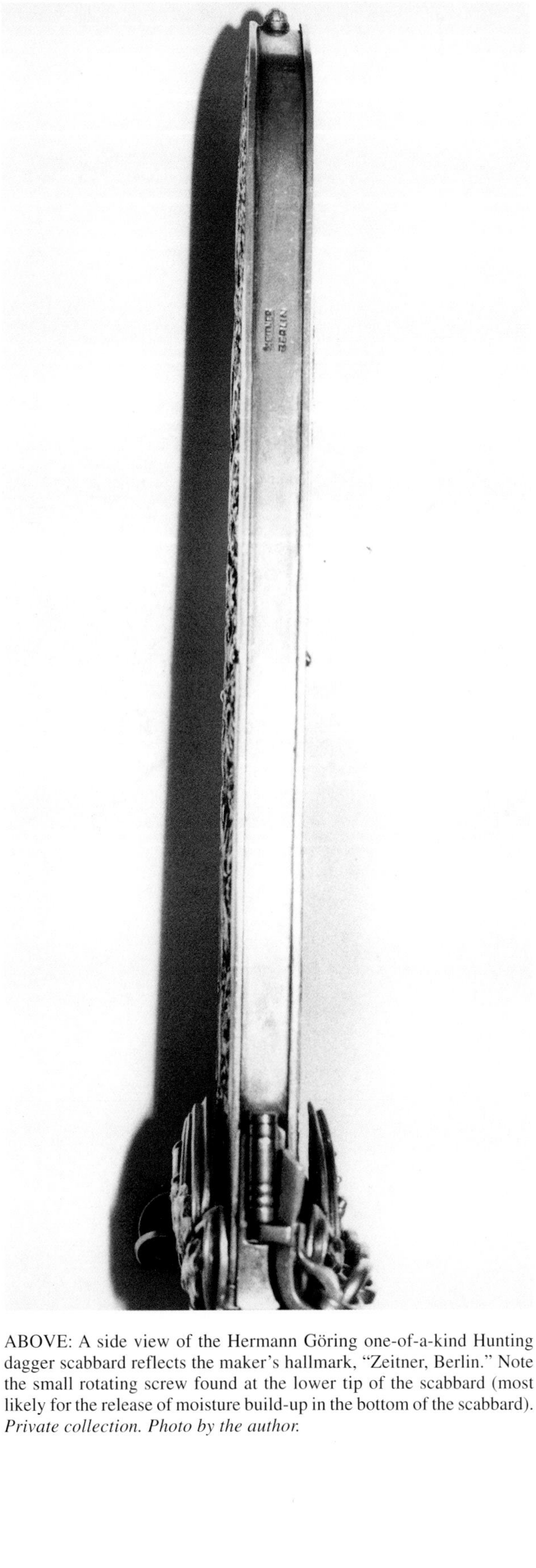

ABOVE: A side view of the Hermann Göring one-of-a-kind Hunting dagger scabbard reflects the maker's hallmark, "Zeitner, Berlin." Note the small rotating screw found at the lower tip of the scabbard (most likely for the release of moisture build-up in the bottom of the scabbard). *Private collection. Photo by the author.*

Göring is pictured wearing one of his two bejeweled Hunting daggers at the *Reich's* meeting of German Foresters at Lake Werbellin. All eyes are on Göring as he inspects an ancient crossbow. *Library of Congress photo.*

As mentioned in the text, Hermann Göring was a flamboyant dresser. The ornate one-of-a-kind dagger presented to Göring by his brother-in-law is shown in this photograph worn with a matching belt. This richly decorated hunting dagger features numerous precious stones and a pure ivory handle. *Photo courtesy of Keith Wilson, Jr.*

The Hermann Göring Bejeweled *Reichsmarschall* Dagger

Ben Swearingen

Author's Note: The author predicted at the close of Chapter I of Volume II of Collecting the Edged Weapons of the Third Reich, "The best bet by this author is that as the hobby of collecting militaria receives more and more international publicity .., more and more rarities (including the Hermann Göring priceless Reichsmarschall dagger) will be unearthed". Obviously, this bold prediction came to fruition during 1994.

Blade collectors of the world may now call off their hunt for the Hermann Göring *Reichsmarschall* dagger. It has been found – at least, what is left of it. All that remains of the most sought-after *Nazi* dagger in the world are a few damaged pieces and a handful of gems.

The story of the fate of the Hermann Göring dagger will dash the hopes of all those who have dreamed of finding and adding this prize to their collection. Initially, I had no desire to write of its destruction. When Tommy Johnson first urged me to do so, I declined. "What's there to tell? It's gone!" However, when I later located the former Commanding Officer of the man who brought the dagger back to the United States, I found myself in the unique position of knowing not only what happened to the dagger after it was brought to the United States – as did several others in the collecting circle – also how the dagger was originally found and disposed of in 1945 in Germany. The latter knowledge only I possessed, and I realized that if I did not share it, the full story of the dagger would probably never be known and recorded.

In the summer of 1945, the 3rd Battalion, 359th Infantry of the 90th Infantry Division, under the command of (then) Captain Orwin C. Talbott moved into the old German training area of Grafenwohr, near Nurnberg. Shortly thereafter, *Frau* Emmy Göring, *Reichsmarschall* Hermann Göring's wife, was transferred from *Schloss Fischhorn* in Austria to *Schloss Veldenstein* in Germany, the castle inherited by her husband in 1938, bringing her into the sector under Captain Talbott's command. "She did nothing to my knowledge in violation of our requirements, and appeared to be almost cowed by her sudden drop in status," recalled Talbott, now a retired Lieutenant General, in a letter to me. "But she was still a cause for concern because of her former high status as, in effect, 'the First Lady of the Third Reich.'"

One of the methods of enforcing the occupation in the first months after V-E Day was sudden and unannounced inspections for weapons and other contraband. In July of 1945, such an inspection was scheduled for the little town in which *Schloss Veldenstein* was located, "not because I was particularly suspicious of her," wrote General Talbott, "but because General Göring might have secreted something there." Though *Frau* Göring had asked for no special privileges, Talbott took care that she should be treated with consideration by the men under his command, and that her personal property would be protected. "In an officers' meeting beforehand, I very specifically ordered that an officer would be present in each room all the while her place was being searched to insure no looting or improper behavior was involved." One particular company under Talbott's command was given the job of conducting the search of the castle. A day or so after the search, *Frau* Göring informed Talbott that much material of importance to her was missing. He immediately drove to the castle, and with the aid of an interpreter became convinced that her complaint was valid. "That night I assembled all officers and some senior non-commissioned officers, except those from the company which had been in *Schloss Veldenstein*. Very shortly after "Taps", we 'raided' their barracks and searched for contraband." Part of the raiding party was stationed outside the barracks, in case contraband was thrown out the windows after the search began. This measure proved to be very wise, as many items were picked up outside the barracks by the waiting officers and NCOs. Numerous other "liberated" souvenirs were found in footlockers.

The next morning, Captain Talbott summoned the Commanding Officer of the guilty company for questioning. "He had not at all obeyed my orders about having an officer in each room as it was being searched. Men were allowed to go to any room they desired." The result was that much looting had taken place, and among the items recovered during the raid was Göring's *Reichsmarschall* dagger.

The dagger presented a problem to Talbott. Unlike most of the other recovered items, it could not be returned to *Frau* Göring, as it was illegal at the time for Germans to possess any weapons. "An alternative I had in mind," recalled Talbott, "was to take the dagger to my commander, and recommend it be forwarded to General Patton as a souvenir 'from' *Reichsmarschall* Göring via the 90th Division." However, for the time being, Talbott retained the dagger in his office.

A day or so after the raid, Lieutenant Wallace Stephenson, one of Talbott's officers about to return to the United States, came to the Captain's office, and asked to be given the recovered dagger. "He said that though for his entire adult life he had been in the Regular Army, he was going to get out, and wanted to buy a chicken farm, but did not have the means. He pleaded to be given the dagger, which he said he could sell and thereby get his farm." Lt. Stephenson's service record had been exemplary, and he had been in combat for two and one-half years and decorated three times for gallantry (Silver Star and a Bronze Star with oak leaf cluster), as well as being the recipient of two Purple Hearts. Talbott realized that this request "was not entirely 'kosher,'" but got "soft-hearted and finally said OK." Talbott admitted to me that he had had "several twinges of conscience over the years for that decision, even if it was not an earth-shaking one."

Talbott heard nothing more from Stephenson until they both attended a 90th Division reunion, perhaps around 1978 or so. Stephenson, "had never been to one before, but had come to this one. He was very friendly, but made no mention of the dagger, nor did I."

It was probably only after I wrote to General Talbott, telling him that Stephenson's son was now in possession of the dagger, that he gave it another thought. "If the son still has the dagger," commented an obviously chagrined Talbott, "it was not sold as Lt. Stephenson had said was going to be the case!"

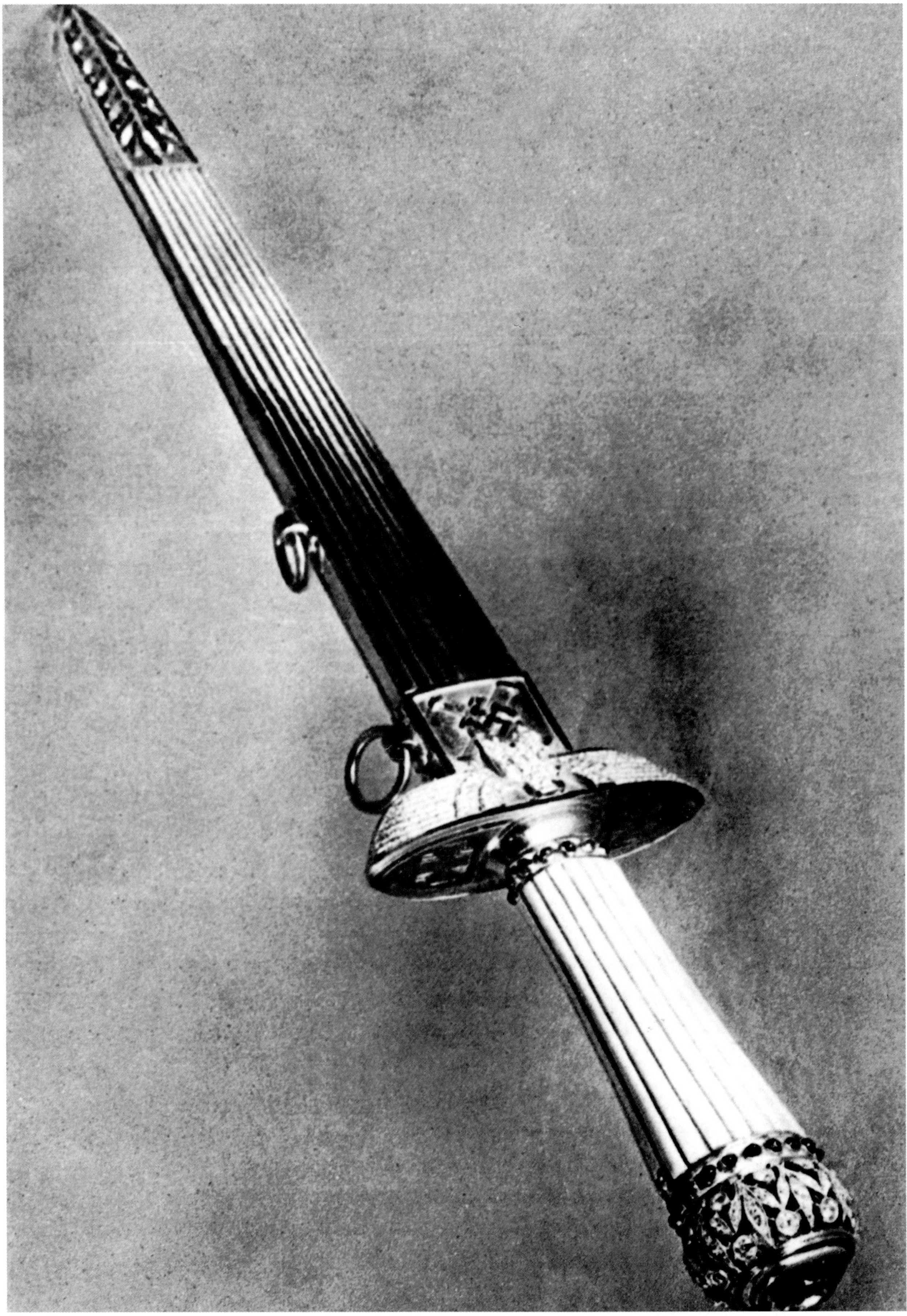

The *Reichsmarschall's* dagger, designed specifically for Hermann Göring by Professor Herbert Zeitner of the Berlin Technical Academy, was manufactured by students there in 1940. The destroyed parts of this magnificent dress dagger recently surfaced in the western United States. *"London Illustrated News" photo library. Photo by T. S. Law.*

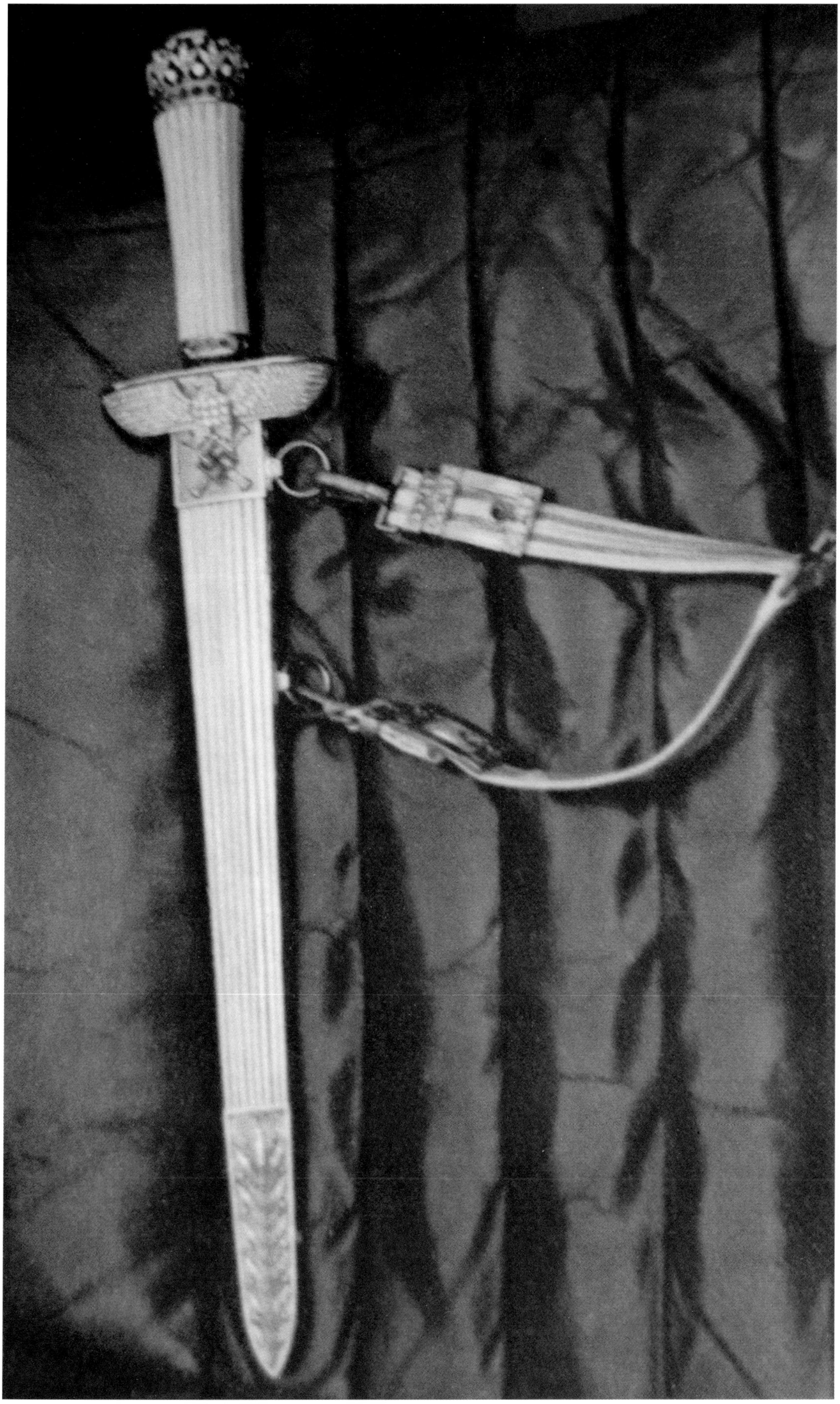

The only other known photograph of the Göring *Reichsmarschall* dagger came to this author in the 1970s after the publication of his first volume of *Collecting the Edged Weapons of the Third Reich*. This historic photograph came via Göring's personal driver and shows the *Reichsmarschall* dagger complete with the accompanying fabric hangers on the rear seat of Göring's Mercedes limousine. *LTC (Ret.) Thomas M. Johnson photo collection.*

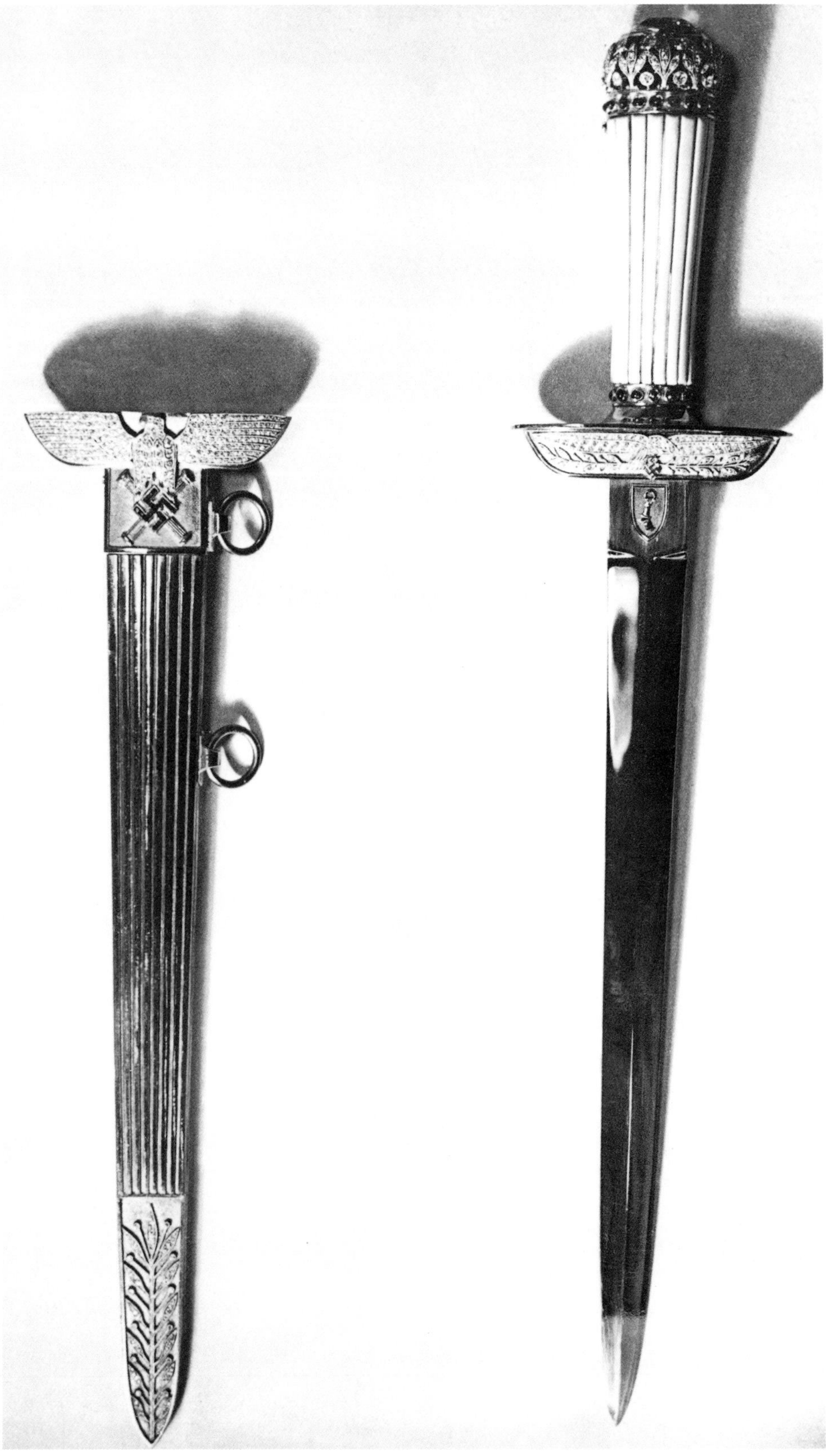

Although Volume VIII of *Collecting the Edged Weapons of the Third Reich* alerted the collecting community of the unfortunate demise of the Hermann Göring *Reichsmarschall* dagger, the missing blade of that one-of-a-kind work-of-art had never been pictured until recently. Speculation has whirled about a possible Paul Müller Turkish Damascus blade or, perhaps, even a plain nickel-plated blade. To serious edged weapon collectors/researchers, this single photograph from Lothar Hartung should be worth the price of this reference. Pictured is the *Reichsmarschall* dagger out of its scabbard. Note the inset Göring "mailed fist" coat of arms on the blade ricasso. This photograph also confirms the information revealed on page 97 of Volume VIII, i.e., the fact that when the dagger was sheathed, its diamond-studded crossguard was completely concealed under the outstretched wings of the *Reichsmarschall* eagle on the scabbard. The author is indebted to historian Andrew Mollo of France for making this extremely rare photograph available. *Lothar Hartung photo collection, Germany. Photo courtesy of Andrew Mollo, France.*

Artist's concept of *Reichsmarschall des Grossdeutschen Reichs* Hermann Göring wearing his one-of-a-kind *Reichsmarschall's* dagger. Göring is shown wearing his diamond-studded *Reichsmarschall's* Baton (the subject of Volume II, Chapter 11 of *Collecting the Edged Weapons of the Third Reich*). *Artwork by Darko Pavlovic, Croatia.*

Lt. Stephenson's son did still have the dagger – again, at least what remained of it – and it was from him that I learned what had happened to the dagger after his father returned to the United States.

Lt. Stephenson evidently had an uneasy conscience about possessing the dagger. Other than to a few members of his family, he mentioned its existence to no one. In effect it "disappeared," leaving behind only a few blurred pictures to excite the imagination of every collector of Third Reich daggers. Only many years later, long after Stephenson's death, did clues to its whereabouts begin to surface in collecting circles. As had been the case several times in the past when matters concerning Hermann Göring were involved, I became an active participant in the unfolding story.

I had just returned from the 1994 Militaria Antiques Xtravaganza (MAX) Show in Pittsburgh, and was sitting at my desk when the telephone rang. A gentleman identified himself as Mack Stephenson, saying he had read my book on Hermann Göring, and asking if he could avail himself of my knowledge on the subject. I agreed, and he began telling me the story of a dagger his father had brought back from the war. (At the time of our conversation, neither Stephenson nor I knew what had actually taken place in 1945, as later reported to me by General Talbott.) He said that his father, an Army officer, had gone to "one of Göring's homes," in 1945, where he found *Frau* Emmy Göring and her daughter in residence. Taking advantage of the opportunity, he asked if there were any weapons in the house. According to the story, *Frau* Göring then handed over to him an ornate and bejeweled dagger, which he eventually brought back to the United States.

From the brief description given my by my caller, I was fairly certain that Mack Stephenson's father had had the incredible good luck to have found the legendary *Reichsmarschall* dagger – and now I had found it! For a moment, my excitement was tempered by the suspicion that I was being "set up." It was too soon after the MAX Show, I thought, to be mere coincidence. However, as I talked with Stephenson, the more I became convinced of his sincerity – particularly after I finally understood that he no longer had the complete dagger, but only a few damaged pieces and the gems! No faker, I reasoned, would employ such a deception. In absolute dismay upon hearing even of the possibility that the *Reichsmarschall* dagger had been destroyed, I refused to identify it until I had more information. I asked Stephenson to sketch the pieces in his possession, and fax the drawings to me. The next morning, after the fax arrived, it was painfully obvious that the mangled pieces had once been part of the *Reichsmarschall* dagger. That moment was probably the most disappointing of my many years of pursuing war souvenirs!

The moment was made worse by the realization that I now had to inform Mack Stephenson that his father had destroyed an item of inestimable historical and monetary value. In a faxed letter accompanying the sketches he had made, he had written happily that he "couldn't possibly express in words my excitement about [the prospect of] finally identifying the object that has held my fascination and awe since childhood." I knew that what I now had to tell him would turn his excitement into dismay. Nevertheless, I wrote to him, telling him what I knew of the *Reichsmarschall* dagger and sent photocopies of pictures of the dagger from the *Collecting the Edged Weapons of the Third Reich* series by LTC (Ret.) Johnson as it originally looked. In a few days, I had his reply. "It is unquestionably the same item," he wrote. "Simply cannot impart to you how I felt after I ripped open the envelope. Jeez ... right there before my eyes, there it was." He handled the bad news much better than I would have done in his place, and only at the end of his letter did his disappointment show through, when he closed, "Thanks for making my day!"

In subsequent telephone conversations and letters, Mack Stephenson related to me the story of what happened to the dagger while it was in his father's possession. His father was evidently in constant fear that some one would try to take it away from him, and once back in the United States during a forty-five-day leave of absence at the end of the war, Stephenson did not even keep the dagger in his own home, but gave it to relatives for safekeeping. "My father kept the object at my deceased uncle's house in Tennessee for many years after the war." According to Stephenson's son, having the dagger in his aunt and uncle's home made them miserable. "That dagger was the reason they had locks on their house, they finally buried it behind their house, and later asked my father to come get his dagger." Once the dagger was back in his possession, Stephenson's fears evidently returned.

"I cannot picture anything other than fear (of what, I don't know) as being the driving force behind his decision to seriously compromise the integrity of the object," Mack Stephenson told me. He went on to relate a very strange incident that occurred during a vacation to another state. "My Dad abruptly picked up and went home when he learned that some 'jeweler in Chicago' was looking for him." What precipitated the jeweler's interest in his father, the son never found out, but he was fairly certain that it had something to do with the Göring dagger. He also believes that it was shortly after that call that the dagger was destroyed.

Stephenson is not sure of the actual date – it may have been in 1961 – that his father made the fateful decision to dismantle the dagger, and extract the jewels. Nor is he sure why his father decided to do so. But for some reason, his father, working with crude tools, and evidently not realizing that the dagger itself had any value, concentrated mainly on removing the gems. All that remains of the *Reichsmarschall* dagger today are the huge aquamarine that once topped the pommel, the damaged crossguard and the crushed throat of the scabbard, the loose diamonds and rubies (or garnets), which had been pried from them, and the diamond-studded body of an eagle. Saving only these few pieces and the loose jewels, Stephenson's father evidently disposed of the rest of the dagger and scabbard. (According to the relative who once had custody of the dagger, and "who probably held the real thing in her hands as many times as Göring had," Mack Stephenson said, the hangers were already missing.)

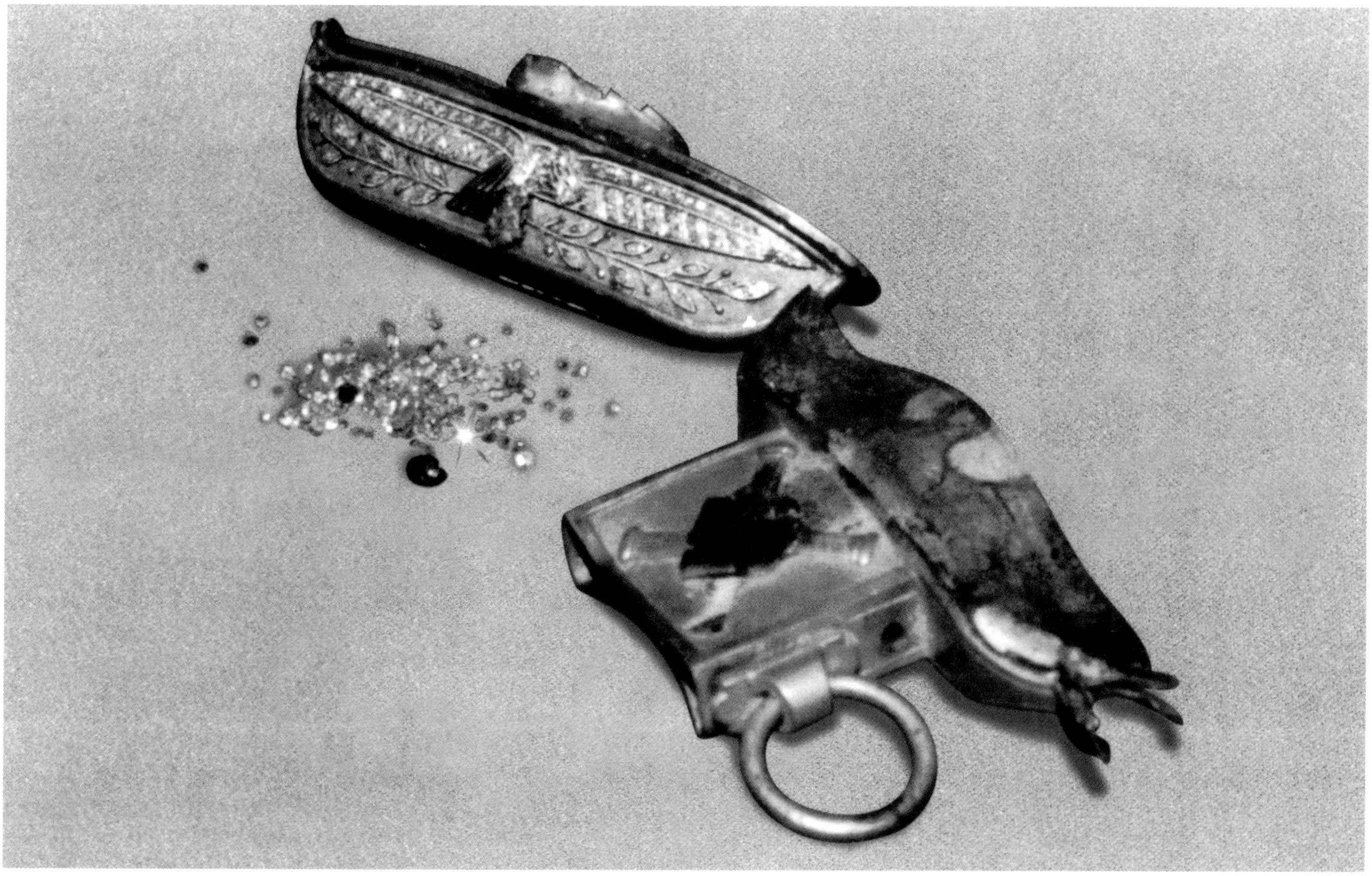

Pictured together are the damaged obverse crossguard on the left and the upper scabbard fitting on the right. Visible between the two are a number of the many precious stones removed from the Göring dagger. Note the severe damage to the upper scabbard fitting. *Private collection. Photo courtesy of the late Robert Waitts and Thomas T. Wittmann.*

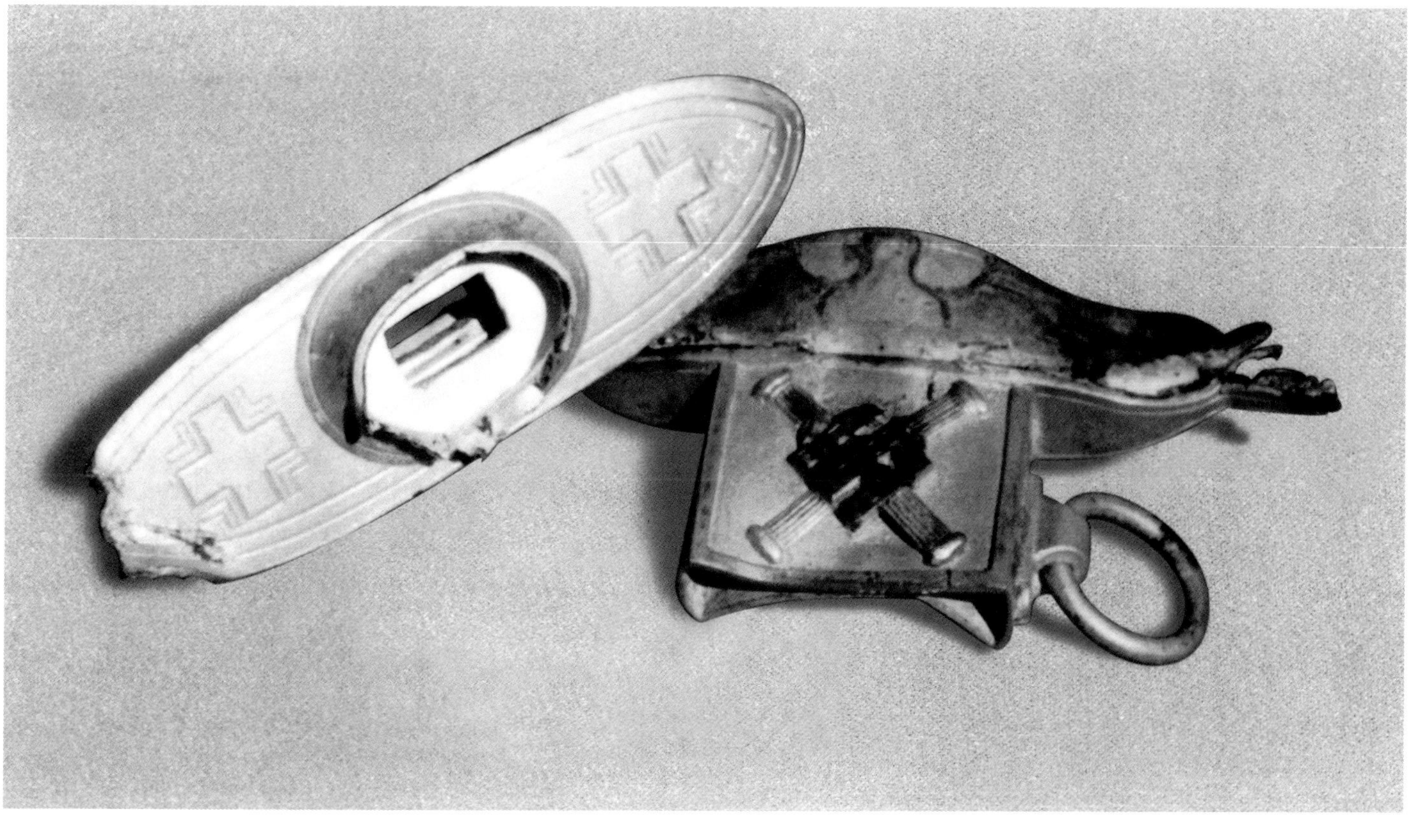

The damaged top of the crossguard and the upper scabbard fitting featuring the *Reichsmarschall* eagle, crossed batons, and swastika. *Private collection. Photo courtesy of the late Robert Waitts and Thomas T. Wittmann.*

Pictured for the second time is the obverse crossguard of the Hermann Göring *Reichsmarschall* dagger. The obverse crossguard features a large, stylized *Luftwaffe* eagle, with its elongated wings and body studded with diamond chips. *Private collection. Photo courtesy of the late Robert Waitts and Thomas T. Wittmann.*

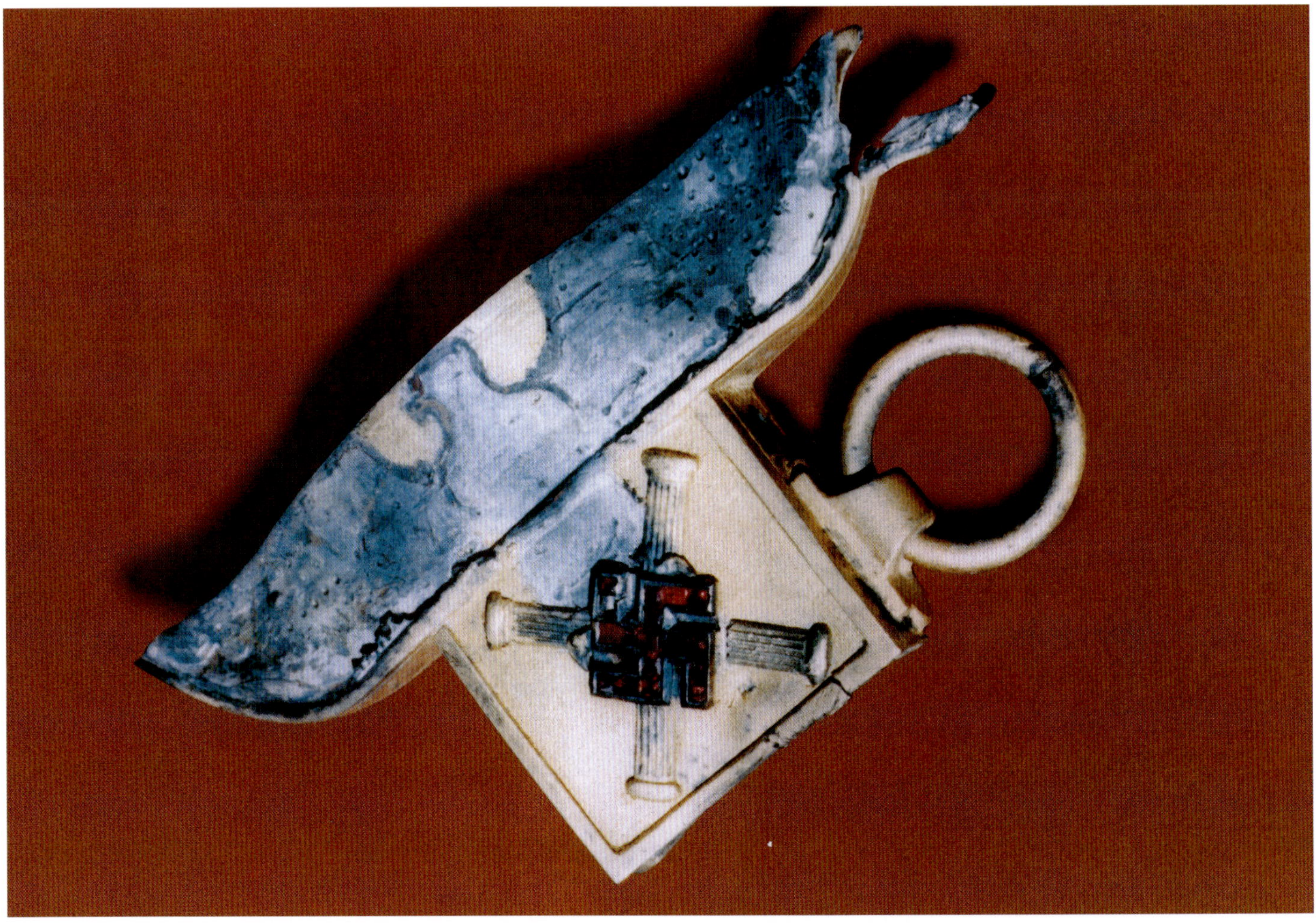

Close-up view of the severely damaged upper scabbard fitting featuring the *Reichsmarschall* eagle, crossed batons, and swastika. Note that all of the diamond chips have been removed from the body and wings of the eagle. The interior crossguard of the *Reichsmarschall's* dagger fits inside the top of the part shown. *Private collection. Photo courtesy of the late Robert Waitts and Thomas T. Wittmann.*

Pictured is the diamond-studded body of the *Reichsmarschall* eagle that has been removed intact from the obverse upper scabbard. An American Roosevelt dime has been included in the photograph for size comparison. Three loose diamond chips are visible to the left of the eagle. To the best of knowledge to this author, none of the remaining parts of the dagger, i.e., pommel, grip, blade, remainder of the scabbard, etc., still exist. *Private collection. Photo courtesy of the late Robert Waitts and Thomas T. Wittmann.*

The majority of the precious stones removed from the Göring *Reichsmarschall* dagger are visible in the photograph of this Ziploc plastic bag. A humble existence to what, undoubtedly, would be the most desirable German Third Reich edged weapon of them all! *Private collection. Photo courtesy of the late Robert Waitts and Thomas T. Wittmann.*

When the entire German edged weapon community was in hot pursuit of the Göring *Reichsmarschall* dagger, the one pictured surfaced in California during 1977 and immediately became one of the enthralling events of the entire hobby. However, shortly thereafter, the California specimen was declared questionable. *The late Roger Steele collection. Photo by Andy Southard, Jr.*

Pictured is the reverse hilt and upper scabbard of the "California *Reichsmarschall* dagger." One of the obvious "red flags" on this specimen is the unexpectedly plain detail of the reverse, which is completely out of character for Third Reich edged weapon design and production. *The late Roger Steele collection. Photo by Andy Southard, Jr.*

A different view of the obverse hilt of the spurious Göring *Reichsmarschall* dagger that surfaced during 1977 in California. Note the jewel-encrusted pommel and crossguard. The top of the crossguard is decorated with two replica Knight's Crosses. *The late Roger Steele collection. Photo by Andy Southard, Jr.*

The obverse blade of the spurious Göring *Reichsmarschall* dagger is engraved, "*Reichsmarschall* Hermann Göring" with a facsimile of Göring's signature. *The late Roger Steele collection. Photo by Andy Southard, Jr.*

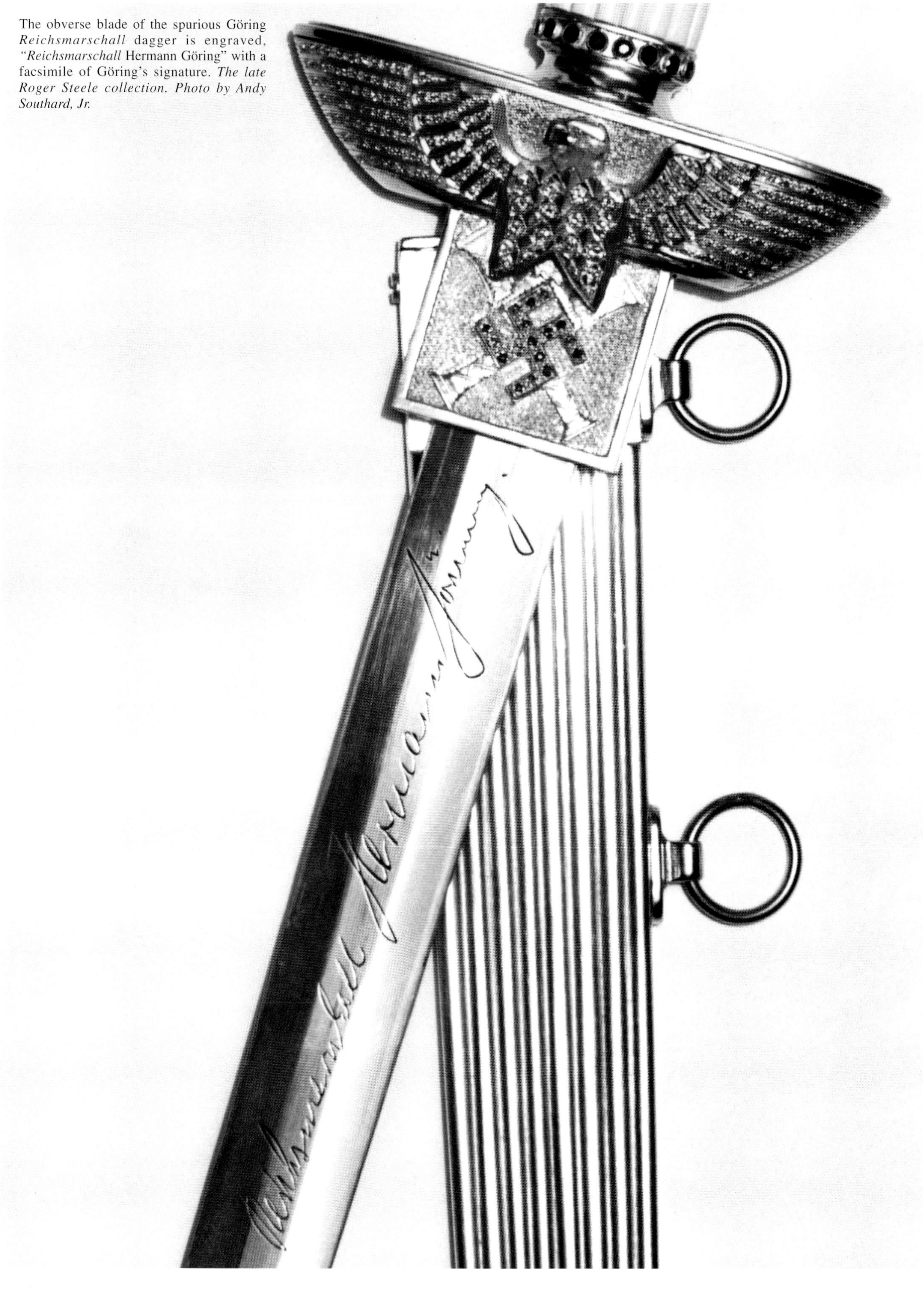

OPPOSITE
LEFT: Reputedly, Hermann Göring's favorite sidearm was not one of the two bejeweled daggers but was a bronze and silver hilted Hunting dagger with a red leather scabbard presented to him by his brother-in-law, Count Eric von Rosen. An inscription engraved on the front of the straight crossguard translates, "A knife from Eric to Hermann." *Photo courtesy of West Point Museum collections.*

RIGHT: The Hermann Göring von Rosen presentation dagger is pictured outside of the scabbard. Göring was wearing this particular sidearm when he was captured and surrendered it to an intelligence officer in the U.S. Seventh Army. The dagger was subsequently donated to the United States Military Academy Museum at West Point. Note the very unusual folding hanging attachment device. *Photo courtesy of West Point Museum Collections.*

ABOVE: This Library of Congress period photograph reinforces the perception that Hermann Göring was an elaborate dresser. Here, he prepares to meet the Yugoslavian Prince Regent and his wife at *Karinhall*. Göring's attire includes his favorite dress dagger. *Library of Congress photo.*

The German Master of the Hunt at *Karinhall.* Note that Göring's hunting regalia included the one-of-a-kind presentation dagger given to him by his brother-in-law, Count Eric von Rosen. *Library of Congress photo.*

Reich Master of the Hunt, Hermann Göring, poses on the steps of his hunting lodge. He holds a plain *Saufeder* (boar hunting spear) in his right hand, while purposely exposing one of his Hunting daggers with his left hand. *Library of Congress photo.*

OPPOSITE: "Life" Magazine published this large black & white photograph of Göring wearing the von Rosen Hunting presentation dagger in their 7 February 1938 edition. Göring is seen escorting Yugoslav Premier Milan Stoyadinovich, who was the first Yugoslav statesman to visit Germany officially since the World War began. *"Life" Magazine photo. Photo courtesy of Harry Logan.*

ABOVE: Hermann Göring, the Supreme Marshal of the German *Luftwaffe* is shown with a group of officials on the grounds of a Hitler Youth training camp for Nazi party leaders located at Ordenburg-Vogelsang. This photo is from a German film seized when the 9th U.S. Infantry captured the former Nazi school. The dress dagger that Göring is wearing presently reposes in the collection of the U.S. Military Academy Museum at West Point. For more information on Hermann Göring and his edged weapons, see Chapter 9 of *Collecting the Edged Weapons of the Third Reich,* Volume III by the author. *U.S. Army Photo.*

At the conclusion of a hunt *Reichsmarschall* Hermann Göring, Master of the German Hunt, discusses the day's events with his fellow hunter. On his left hip he wears the custom-made dagger presented to him by his brother-in-law, Count von Rosen. The bronze and silver hilt and scabbard fittings are enhanced by the bright red leather scabbard covering. Göring was wearing this one-of-a-kind dagger at the time of his surrender in May 1945. *LTC (Ret.) Thomas M. Johnson photo collection.*

In this Heinrich Hoffmann photograph of a portrait of Hermann Göring as *Reichsjägermeister*, he wears his one-of-a-kind presentation dagger given to him by his brother-in-law, Count von Rosen. The dagger features a stiletto blade with hilt and scabbard fittings constructed of bronze with the scabbard covered in red leather. This rare dagger is now on display in the United States Military Academy at West Point, New York. *LTC (Ret.) Thomas M. Johnson photo collection.*

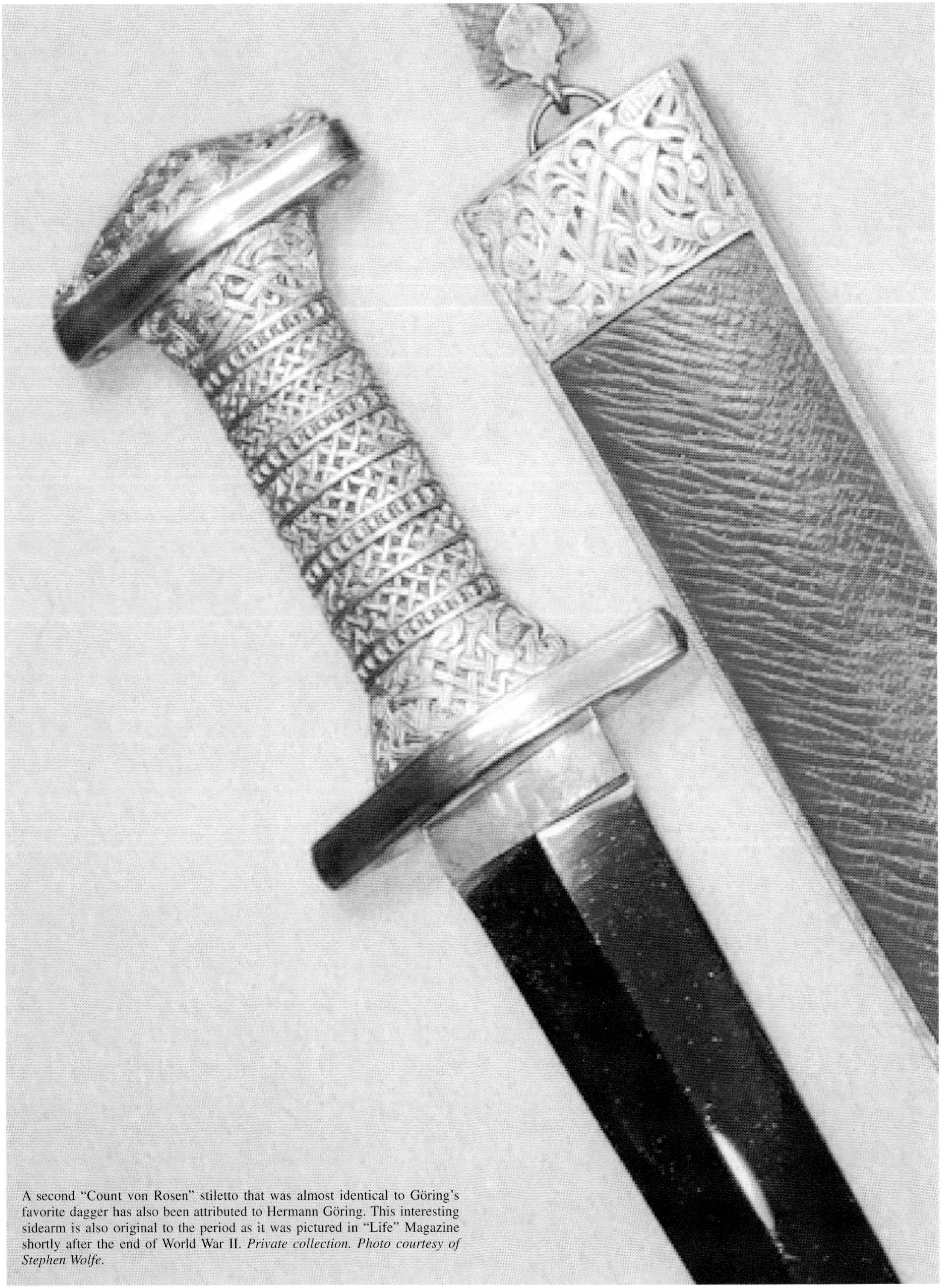

A second "Count von Rosen" stiletto that was almost identical to Göring's favorite dagger has also been attributed to Hermann Göring. This interesting sidearm is also original to the period as it was pictured in "Life" Magazine shortly after the end of World War II. *Private collection. Photo courtesy of Stephen Wolfe.*

Another spectacular dress dagger handmade for Hermann Göring but, evidently, never presented, was the massive gold Göring Industrial dagger. This unique piece was discovered by the late LTC Jim Atwood at the Alcoso firm in Solingen, circa 1960. The blade had been broken, and Atwood commissioned Master Sword Smith Paul Müller to forge a matching replacement blade. The hand carved oak case is not original to the period. *Private collection. Photo by Andy Southard, Jr.*

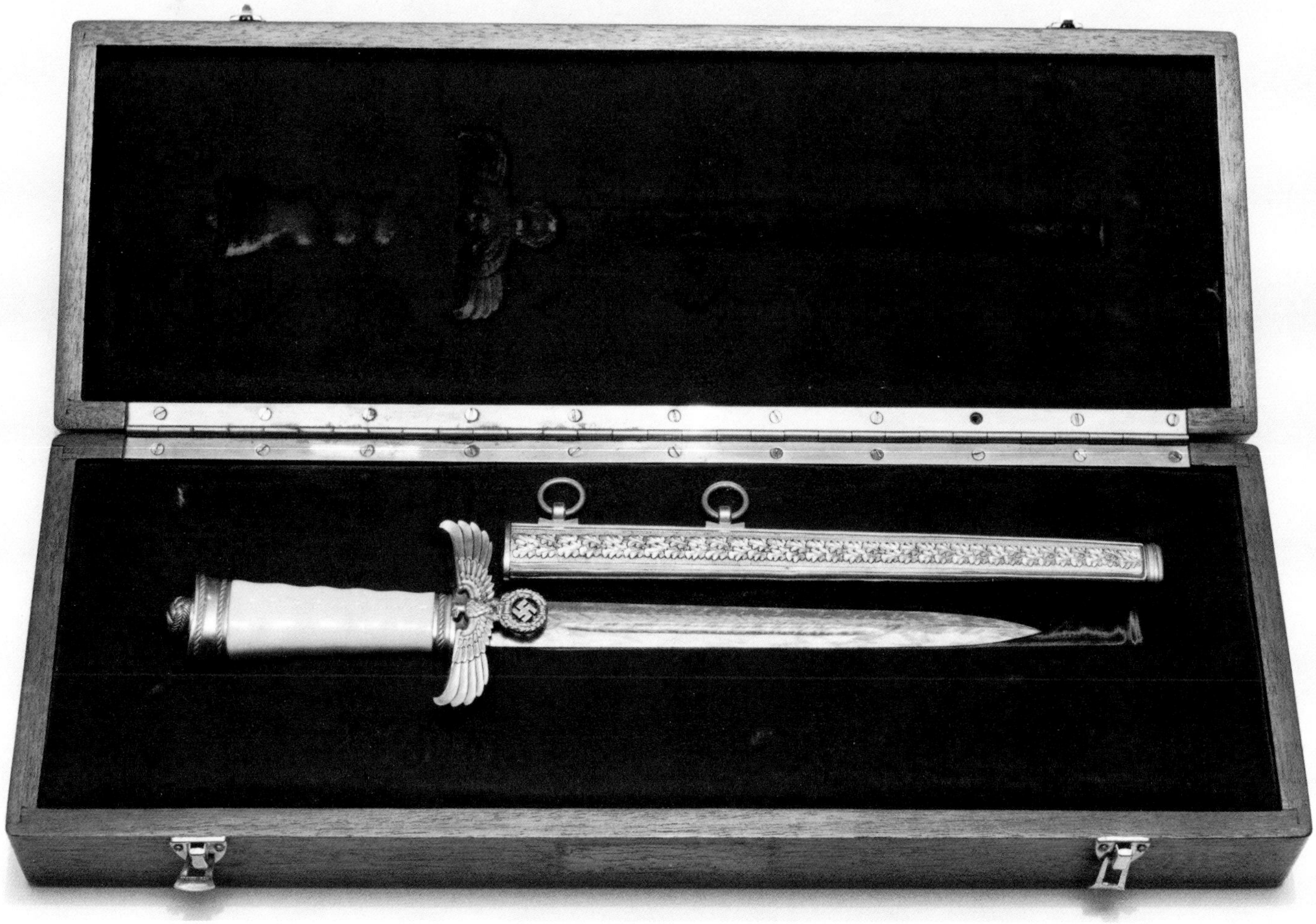

Interior view of the postwar oak display case that houses the one-of-a-kind Hermann Göring Industrial dagger. *Private collection. Photo by Andy Southard, Jr.*

CHAPTER TWENTY-FIVE

Reproduction Third Reich Edged Weapons and How to Spot Them

"If all mankind were suddenly to practice honesty, many thousands of people would be sure to starve."
— G.C. Lichenberg

The response to an early request for suggested subject matter for this reference was gratifying. Siphoning through the plethora of requests and comments, one subject stood out – reproduction *Nazi* edged weapons and how to spot them! Any collector who has undergone the unfortunate experience of spending hard-earned dollars for a "super rare" Third Reich edged weapon which subsequently turned out to be a phony is understandably anxious to preclude making the same mistake again. Unfortunately, unscrupulous dealers and collectors abound and very few sizeable collections are completely void of any phony pieces.

This chapter will not delve into the specific reproduction mistakes made for each particular model Third Reich edged weapon. This subject area is vast and warrants a separate reference book in itself. Three preliminary efforts dealing specifically with reproduction *Nazi* edged weapon models are a paperback pamphlet published in England by R. McFarlane in July, 1969, entitled, *"Bluebook of Identification of Reproduction Nazi Edged Weapons"*, a section devoted to the subject in John R. Angolia's book on edged weapons entitled, *"Edged Weaponry of the Third Reich"*, and a separate expanded reference work on the subject of reproductions entitled, *"Reproduction? Recognition!"* by British author Frederick J. Stephens.

The theme of this chapter will be to provide the researcher/collector with some general guidelines that will serve him well in attempting to identify counterfeit pieces. Albeit, there is absolutely no substitute for experience in examining edged weapons, there do exist several reproduction indicators that will serve even the beginner well. Usually the individual with the inherent facility to never be "taken" by a reproduction edged weapon is the same individual who has been an ardent collector and researcher for many years. The correlation between years of experience and probability of being duped is obviously not due to chance alone. The reproduction indicators discussed below will be referred to as "red flags" because, hopefully, each will serve to mentally raise a red flag in the collector's mind prior to engaging in a bad purchase. Once one or more "red flags" listed below have been identified, it is recommended that the particular piece in question not be purchased. Although not necessarily a certainty, the item is more than likely a reproduction. In this field of collecting, it is much better to be too cautious than not cautious enough!

1. **New Appearance.** The vast majority of edged weapon reproductions on the market today appear to have been manufactured within the past few weeks (and might well have been!). Quite frankly, 65+ years of age will tend to leave some telltale signs, regardless of the care and storage means utilized. When examining a piece, look at the screw heads. Are they bright, shiny, and completely free of any dirt or corrosion? Inspect the small leather washer under the crossguard (if one is available). If it appears as new as the leather on a belt that you just purchased, proceed with caution. Take a flashlight and have a close look at the *inside* of the scabbard throat. If all the internal parts appear to have been made earlier in the month, they probably were! Unfortunately, this new appearance indicator will not stand alone, as some unscrupulous individuals have discovered such devious means as burying, soaking in urine, etc., to purposely "age" reproduction pieces. Keep in mind that old appearance alone is no guarantee of authenticity.

2. **Misfitting Parts.** To state that German quality control is usually superior would probably be the understatement of the year. The meticulous quality control exercised by the Germans in the manufacture of automobiles, cameras, etc., is held in esteem the world over. The fact that Third Reich weapon quality control was superior is evidenced by the recent reproductions manufactured in the United States, Spain, China, England, etc., being no match for the originals. Thus, be wary of *any* misfitting parts. For example, if an *SA* dagger wooden grip bears large gaps between the handle and the crossguard, rest assured that it never would have left the factory. Likewise, an *SA* eagle and swastika grip insignia that rests in an indentation much too large to accommodate it should definitely raise a "red flag". In general, be on the lookout for crudeness in manufacture and/or fit of component parts. Also take the time to include exact dagger dimensions in your edged weapon reference library. Numerous reproductions have grossly inaccurate dimensions. For example, the postwar reproduction of the ultra

rare Diplomatic bayonet made in England is significantly larger than the original bayonet.

3. **Unmarked Interior Parts.** If the edged weapon that you are examining lends itself for disassembly, carefully take it apart and scrutinize the interior component parts. Take the time to learn what markings, if any, should be evident. For example, do you realize that the inside of both TENO EM grips should feature the familiar Eickhorn squirrel trademark? The tangs of several different model dagger blades were carefully marked with the manufacturer's initials and/or mark. Often each component part was stamped with a corresponding serial number vis-à-vis the Diplomatic and Government Official's daggers. Take the time to learn what to look for when you disassemble a particular model sidearm.

4. **Unusual Variations.** With the ever-increasing escalation of Third Reich edged weapon prices, a multitude of "one-of-a-kind prototypes" are finding their way into the market. While some of these pieces are indeed authentic prototypes, a "non-documented" prototype should be approached with a *great* deal of caution. The vast majority of "prototype" *Nazi* daggers being offered for sale today are reproductions, and since the prototype pieces demand top dollar, insist upon complete documentation prior to the purchase of one. Fortunately, the majority of the various Solingen *Waffenfabriken* sales catalogs survived the war and represent a collector's primary reference source. A tedious search of all reference sources available should be made prior to the outlay of huge sums of money for unusual variations or prototype pieces.

5. **Faulty Engraving.** Like the previous indicator, the premium prices sought for Third Reich edged weapons bearing engraved blades have opened the flood gates for reproduction engraved bayonets, daggers, and swords. Not only are complete blades currently being manufactured, but unscrupulous dealers and collectors are resorting to having the local jeweler engrave crossguards, scabbard fittings, etc., with fictitious (and sometimes actual) German names and units. Advice on the engraving issue would be two-fold: first, study the engraving style, depth, etc., of known original pieces. More times than not, the local jeweler's version is completely "foreign" to the characteristic German patterns utilized during the war years. Second, resort to the original *Waffenfabriken* sales catalogs to ascertain if a particular standard engraved blade pattern was in fact a product of the manufacturing firm in question.

6. **Incorrect Proofmarks.** The size of the edged weapon factories during World War II ranged from mammoth corporations to small "cottage-craft" shops operated in the rear of Solingen homes. Obviously, each individual firm did not manufacture the entire plethoric gamut of Third Reich sidearms. In some cases, a single firm designed a particular model dagger and application for a patent was made. The blade was then stamped with the *"Ges.Gesch."* (Patent Pending) mark. Prime examples of patented model designs are the TENO Officer and Enlisted daggers by the Carl Eickhorn firm. Even the more common models were often restricted to several selected manufacturers. Armed with extensive knowledge of which firms produced which sidearms, the wise collector can rapidly eliminate a number of phony pieces bearing incorrect proofmarks. For example, how many of the readers of this chapter could accurately consolidate a list of the *only* Solingen factories to produce Naval daggers for the Reich? The completed list should look like this – Alcoso, Clemen *und* Jung, Eickhorn, Höller, Hörster, Robert Klaas, Krebs, Lauterjung, Luneschloss, Pack, Plumacher, Puma, Paul Seilheimer, SMF, Emil Voos, Max Weyersberg, Paul Weyersberg, Wingen, and WKC.

7. **Incorrect Accouterments.** An excellent "red flag" source is the accompanying accouterments to a particular sidearm. Although it is obvious that hangers, frogs, and knots are interchangeable and are often switched on authentic pieces, the reverse is usually true with reproductions. Most "repros" are manufactured complete to include the accompanying leather or fabric accouterments. Thus, the wary collector is provided with yet another invaluable red flag source. When examining accouterments, make a careful inspection of the inside of leather items. Does the natural leather and thread stitching exhibit 65+ years of aging? If all of the known original standard bayonet frogs that you have observed were constructed of smooth leather, and you are offered the "opportunity" to purchase one constructed of pebbled leather – BEWARE. Check the condition and wear of the portepee/knot. Does it appear to be recently manufactured? One final word of caution – some unscrupulous dealers will add authentic trappings to a reproduction sidearm in order to avoid this particular "red flag". Thus, one should not attempt to let this indicator stand alone, but utilize it in conjunction with the previously mentioned indicators.

8. **Incorrect Factory Markings.** Should you be fortunate enough to acquire a factory new (unissued) piece, compare the manufacturer name listed on a cardboard issue tag and/or paper shipping bag with the *RZM* code or proofmark engraved on the blade. Obviously, if the piece is unaltered, the manufacturer name listed on the tag and shipping bag should be the same company whose *RZM* number or proofmark appears on the blade. *SA* daggers have been observed recently having WMW *(Waffenfabrik Max Weyersberg)* paper issue tags affixed to the upper scabbard fitting ring and *RZM* numbers of manufacturers other than WMW on the blades.

9. **Non-Existent Models.** Unscrupulous dealers have gone as far as to promote a demand for "original" Third Reich blades that never even existed under the Reich! The best example of this fraudulent effort is the brass Eickhorn *Schutzstaffel (SS)* pocketknife that, unfortunately, made its entry into the marketplace several years ago. These spurious *SS* knives are presently being manufactured in England and are 100% frauds. Interestingly enough, this particular fake has been manufactured with built-in aging and appears to be original in all respects. However, as mentioned in the discussion of the first indicator above, "aging" can be accomplished by artificial means. The wholesale price of these knives from the manufacturer is only a few dollars each. However, since research indicates that this piece never existed and is a

complete fraud, its true value is much less than even the wholesale price. As a matter of fact, strictly from a collector's standpoint, this item is worthless and only tends to mar an authentic collection of Third Reich edged weapons. Hardly a week goes by that some unsuspecting collector does not contact the author "excited" about locating one of these "rare" sidearms.

In conclusion, the adoption of three general "rules of thumb" for collecting Third Reich edged weapons is recommended. First and foremost, become a *student* of the subject. Do not rely on gun show talk and the opinions of other collectors to educate oneself. *Learn the facts for yourself.* The best way to begin is to build oneself a large reference library on the subject. The limited amount of published works in this field will preclude the price of a fine library from becoming prohibitive. Many collectors have made the unwise statement, "I would much prefer to allocate the $50.00 price tag of a current reference book toward the purchase of a good dagger." The truth of the matter is that the $50.00 expended for a good reference book might repeatedly preclude throwing away sizeable sums of money on *bad* daggers. The famous quote by Benjamin Franklin is indeed apropos, e.g., *"If a man empties his purse into his head, no man can take interest."* Attempt to augment your current reference books with as many original Solingen sales catalogs as you can obtain. These catalogs have the distinct advantage of depicting *only* original materials, whereas a few reproduction pieces, unfortunately, grace the pages of most of the current reference books.[11] Copies of the original catalogs and reprints of several catalogs are available from various dealers.

The second rule of thumb pertains to the subject of dealers (or sources). Find one or two whom you can trust explicitly and direct your total dealings their way. The integrity of Third Reich edged weapon dealers (and collectors) in this country and abroad runs the gamut from beyond reproach to totally unscrupulous. Fortunately, the hobby is small enough that a few fast inquiries to other collectors will usually distinguish the dealers and collectors to avoid. Those dealers or collectors who have sold reproductions for genuine pieces will be *rapidly* identified!

A third general rule of thumb that will serve you well is if you have *any* reservations about the authenticity of a particular edged weapon, leave it alone. Psychologists refer to this uncertainty discomfort as "cognitive dissonance". Obviously, the dissonance may occur before or after the purchase, but much better for the purchaser if the dissonance is initiated early by one or more "red flags", thus negating the purchase. An honest collector will readily admit that the vast majority of his "maybes" turned out to be reproductions. Conversely, if you should see an item for sale at a show, in a shop, or on a dealer's list, that appears to be original and would fill a sought-after hole in your collection, do not procrastinate. A common sign to be found in antique shops is, "The best time to buy an antique is NOW. If you wait, it will be gone!" The same rationale is certainly true of World War II edged weapons.

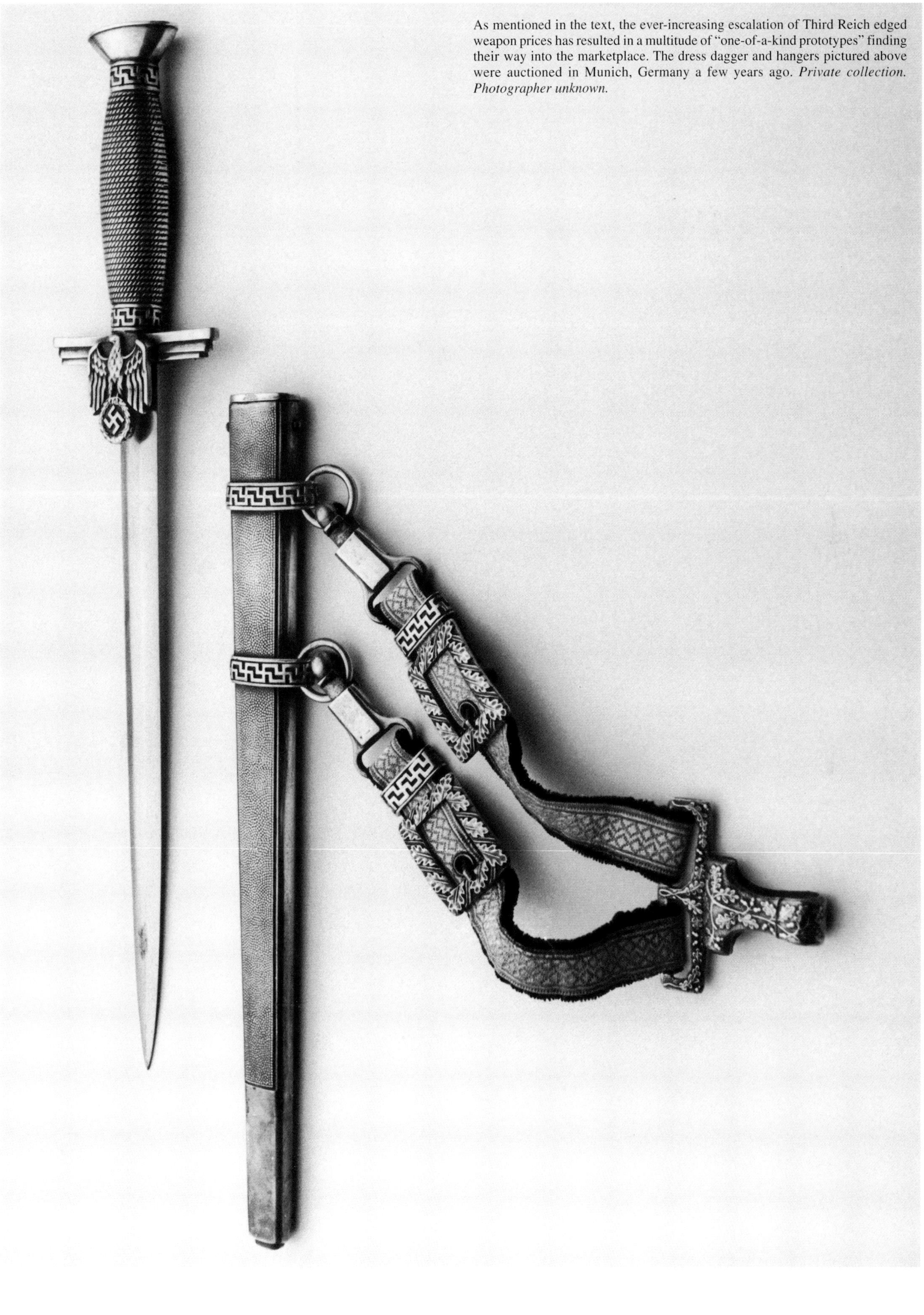

As mentioned in the text, the ever-increasing escalation of Third Reich edged weapon prices has resulted in a multitude of "one-of-a-kind prototypes" finding their way into the marketplace. The dress dagger and hangers pictured above were auctioned in Munich, Germany a few years ago. *Private collection. Photographer unknown.*

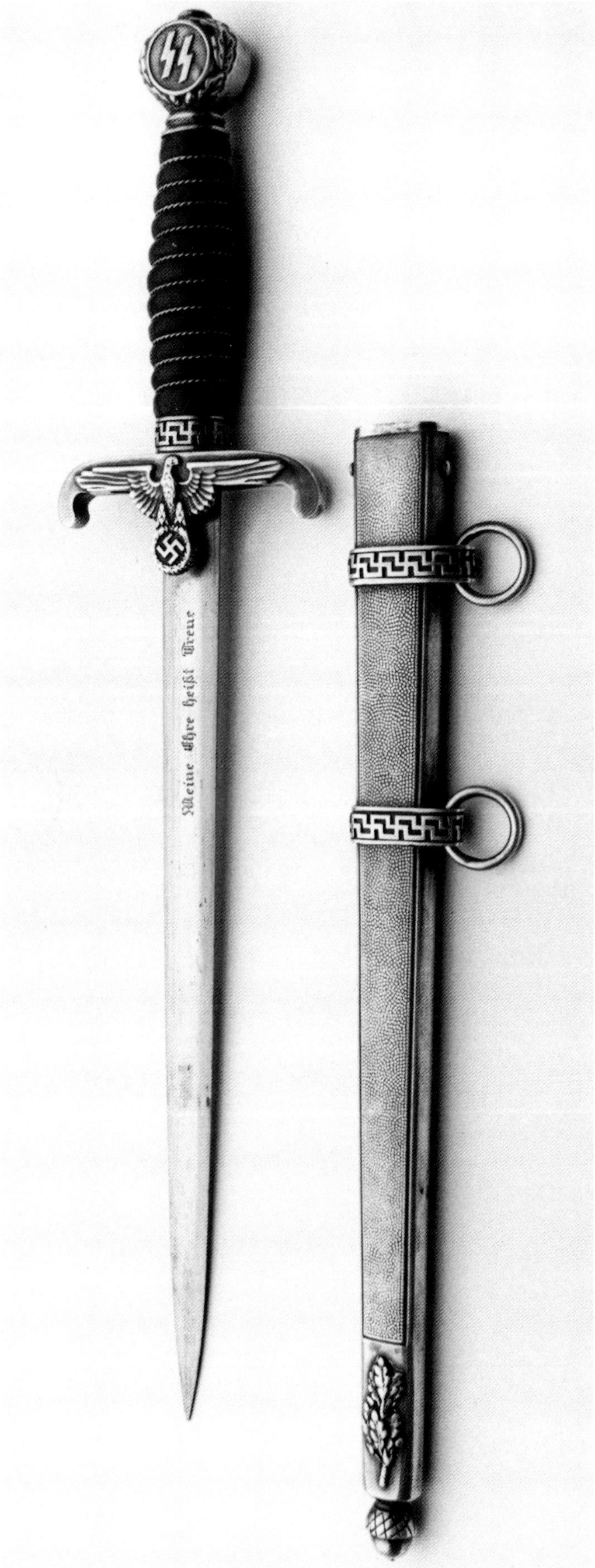

Pictured is another "fantasy" Third Reich dress dagger that was also offered to the collecting community during the 1970s. This was supposedly a rare "one-of-a-kind" SS prototype dagger. *Private collection. Photographer unknown.*

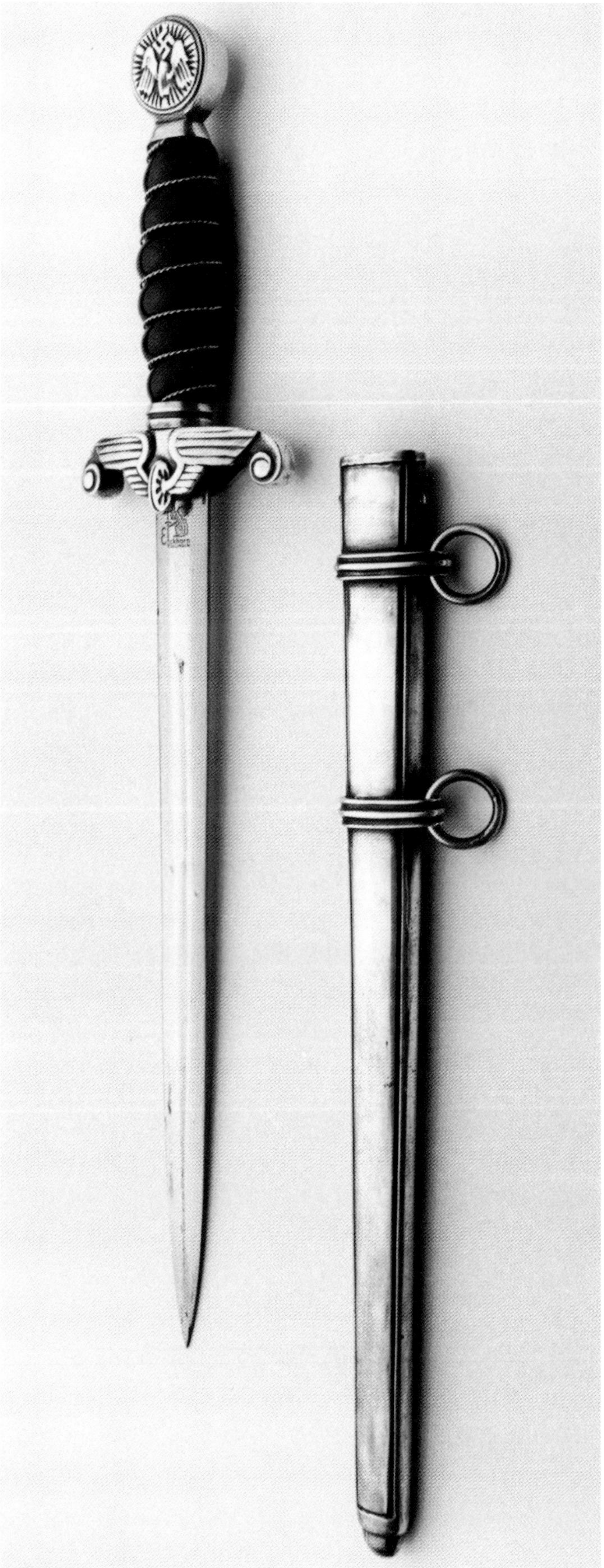

Pictured is a Third Reich *Bahnschutz* (Railway) fantasy dress dagger. Note the incorporation of a proper Carl Eickhorn "squirrel holding sword" (1935-1941) trademark on the obverse blade ricasso. The quality of many reproduction daggers is exceptional. *Private collection. Photographer unknown.*

A currently made Army Field Marshal's dagger whose prototype design was technically in existence prior to the close of World War II. The original design was published by the industry and trade union of Solingen along with the proposed design for a Field Marshal's sword. No known original examples of either model were ever produced. *Thomas W. Pooler collection. Photo by Andy Southard, Jr.*

Close-up of the currently-made Army Field Marshal's dagger. *Private collection.*

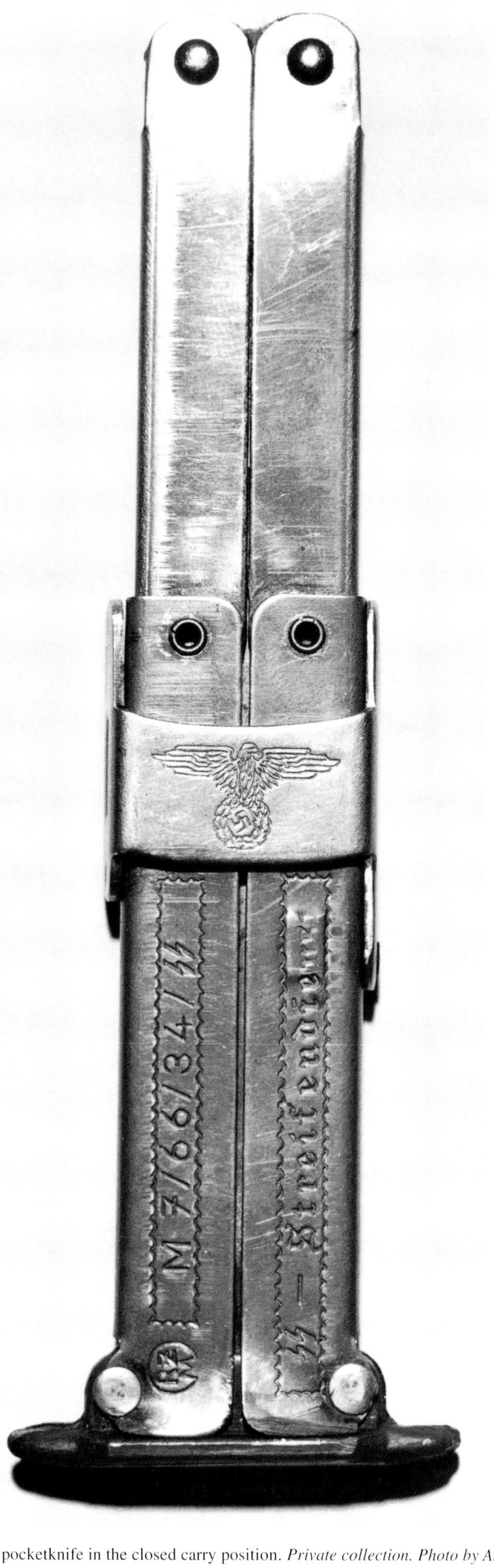

The fake SS folding pocketknife in the closed carry position. *Private collection. Photo by Andy Southard, Jr.*

The pictured SS folding pocketknife is actually a fake as opposed to a reproduction, as it never existed. Manufactured in England, these knives are even "aged" in advance to appear original. Shown here is the obverse view in the half-extended position. *Private collection. Photo by Andy Southard, Jr.*

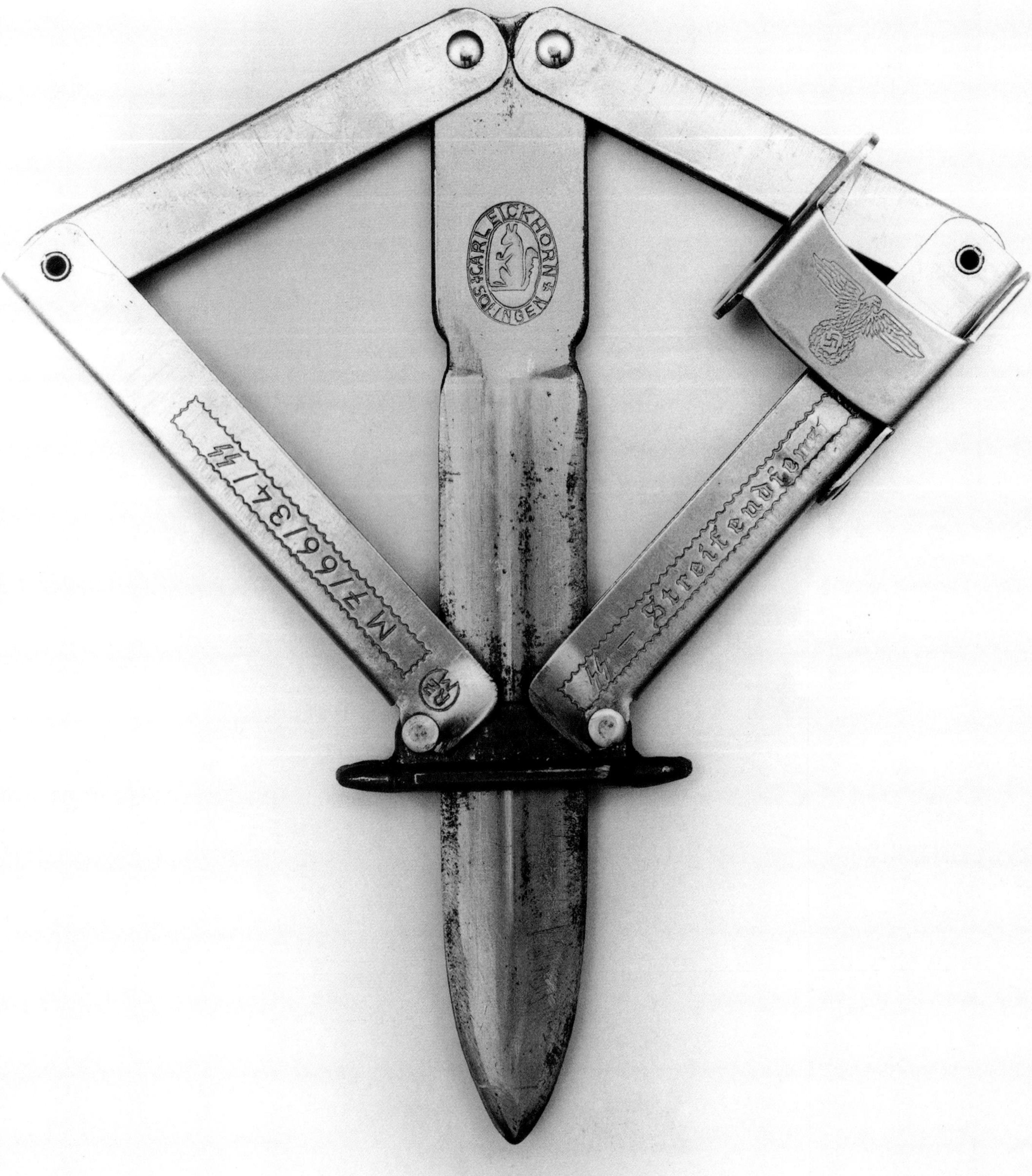

The spurious Eickhorn SS folding pocketknife in the half-open position – reverse view. *Private collection Photo by Andy Southard, Jr.*

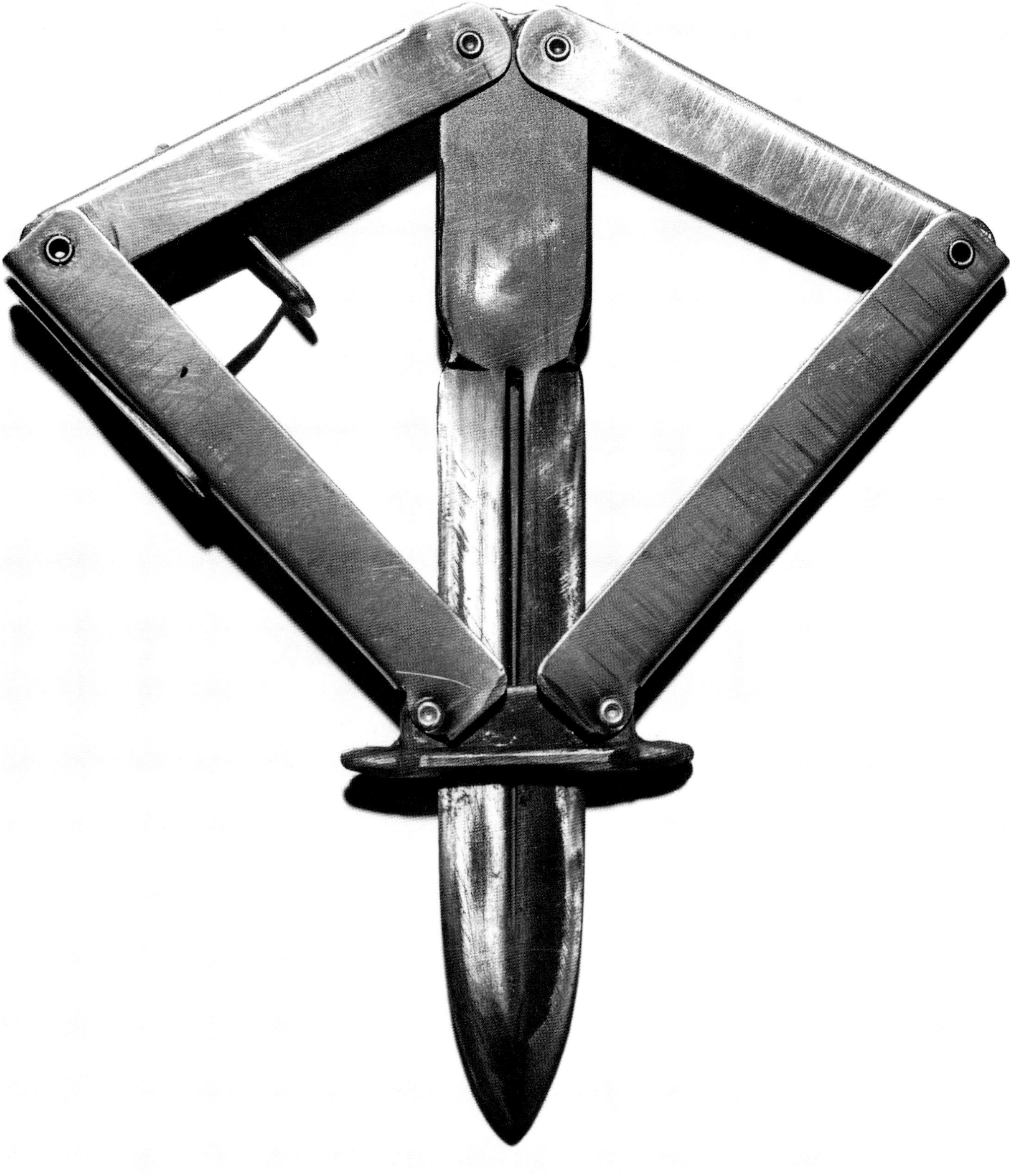

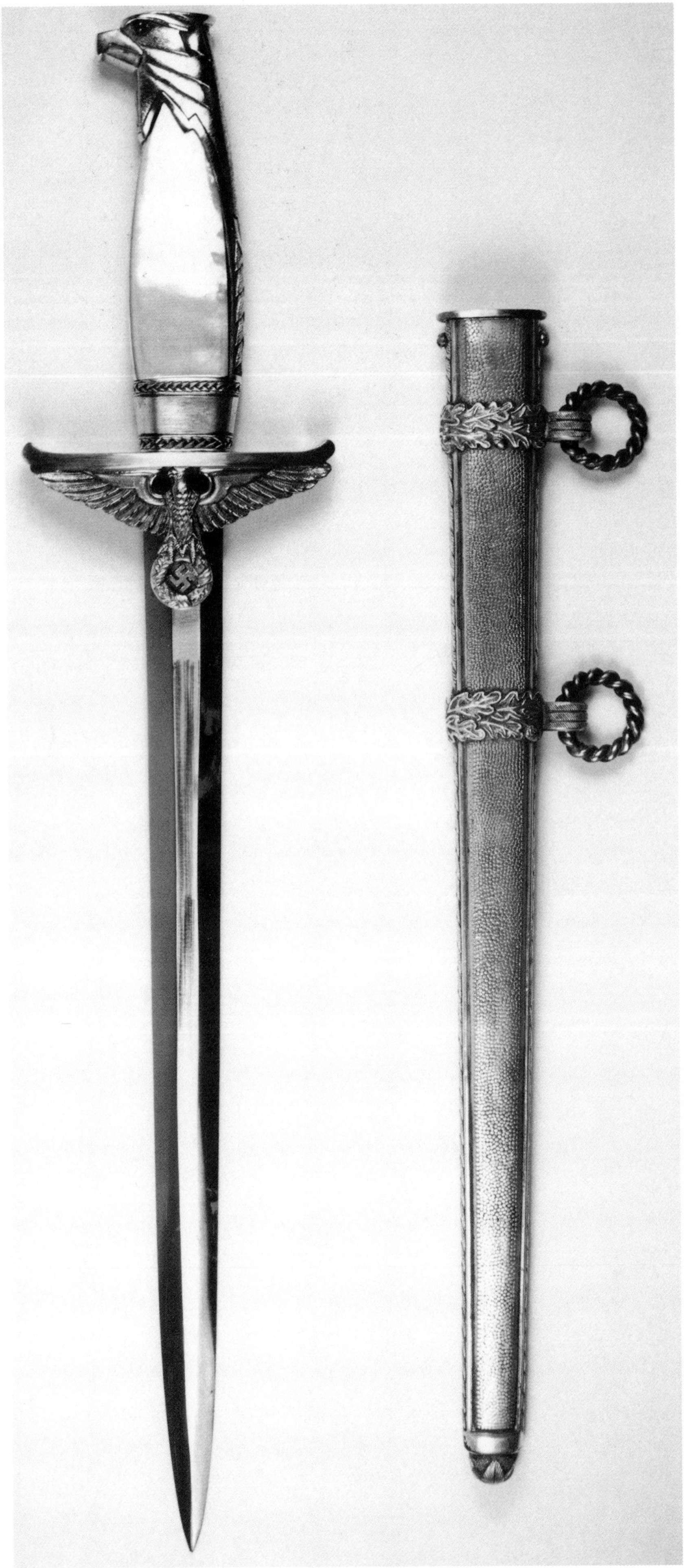

LEFT: A custom-made reproduction of a Government Official's dagger in silver – obverse view. Note the intricate detail and craftsmanship that went into this dagger's manufacture. Note also that the eagle head pommel has jeweled eyes. *Private collection. Photo by David Delich.*

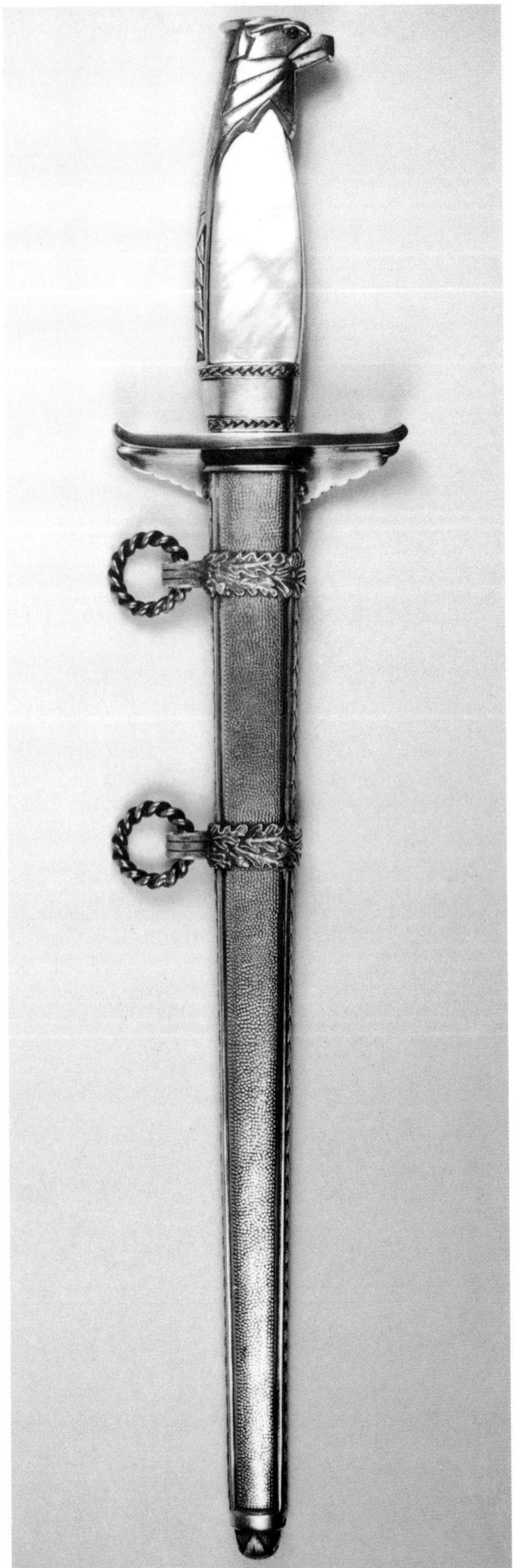

ABOVE: Custom-made reproduction Government Official's dagger – reverse view. *Private collection. Photo by David Delich.*

Close-up view of the obverse crossguard of the custom-made reproduction Government Official's dress dagger. Note the intricate detail of the crossguard, ferrule, backstrap, and scabbard ring. This single specimen is, undoubtedly, one of the finest quality reproduction German daggers made to date. *Private collection. Photo by David Delich.*

Current Solingen Production

Solingen, today, remains a thriving city with a population of approximately 160,000. Largely rebuilt following the heavy Allied bombings in World War II, the "City of Swords" continues to relate to fine cutlery as Paris relates to fashions, Detroit to automobiles, and Amsterdam to tulip bulbs. As during the war years, cutlery manufacturing firms run the gamut from large automated plants and mills to the minuscule "cottage industries", where items are made by artisans in their own homes.

Many collectors are not aware of the fact that a few of the standard model sidearms carried by members of Hitler's Reich are not unique to the period 1933-1945. Some of these models were being manufactured and issued prior to the Third Reich, and more important yet, some are still being manufactured in Solingen today! Needless to say, reference is made to the edged weapon models which were completely void of the swastika symbol.

Some of these remaining firms still manufacture selected daggers, bayonets, and swords that were in use during World War II. It should be pointed out that these pieces are not being manufactured as reproductions with an intent to deceive the collector market, but instead are production items manufactured for current consumption as edged weapons continue to be issued in Germany to members of the Navy and Forestry Service. In most cases, the *Waffenfabriken* order numbers even remain the same as during the 1933-1945 period. For example, the Carl Eickhorn firm before their 1976 bankruptcy manufactured and sold the identical Army Lion Head Cavalry sword, model No.1312, that is illustrated in the original Eickhorn *"Kundendienst"*. Likewise, the unadorned Army saber, model No.189, was offered by the identical number. In the dagger line, the Fire Official's dagger, model No.42, and numerous Forestry and Hunting pieces illustrated in the "*Kundendienst*" could be obtained long after 1945.[12] A January, 1975, letter from the Eickhorn firm to the author states that the gold-plated Water Protection Police Leader's dagger, model No.1797, has been placed back into production as a replica and is currently available for 130 DM. The collector should realize that, with aging, these pieces are practically, if not indeed, impossible to distinguish from the original World War II models. The majority of the Solingen firms will make available upon request their current sales catalogs, brochures, and price lists. The serious collector should be well aware of which model edged weapons are presently being produced and by which firm(s).

The entirety of the Solingen postwar blade production has not been as innocent as described above with the current production pieces. No sooner had the war ended until profit-minded Solingen plant owners took note of the souvenir desirability of the Reich edged weapons to the victorious Allied occupation troops and began assembling standard *SA* daggers from original and, if unavailable, postwar production parts. In one instance, the Carl Eickhorn firm was clever enough to market leftover 1934 Pattern *Luftwaffe* daggers to U.S. forces by modifying the blades with English inscriptions, "Field Day" and "Germany 1945". These pieces were evidently utilized as prizes at an American military field day.

Some observations are in order at this point concerning the conflicting usages and connotations of such terms as, "new", "original", "original parts", "parts dagger", "reproduction", and "replica". Although the definitions listed below are strictly the author's, they should prove helpful to the collector/researcher in distinguishing between misconstrued terms.

1. **Factory new (unissued) –** 100% original finish and parts, everything immaculate, i.e., numerous Allied soldiers entered the Solingen manufacturing firms at the termination of World War II and "liberated" edged weapons off the shelves. Many of these pieces come complete with original *RZM* issue tags and factory shipping bags. Extremely rare.

2. **Original –** manufactured and assembled during the period 1933-1945.

3. **Original parts –** assembled from stocks of original parts remaining in the hands of the original manufacturers when World War II ended. Must consist of 100% original parts.

4. **Parts –** a composite of original and postwar manufactured parts.

5. **Reproduction –** a spurious edged weapon manufactured post-1945 and consisting of 100% postwar manufactured parts.

6. **Replica –** same as reproduction.

The serious collector can save a lot of grief by developing a personal checklist on each type of edged weapon that he anticipates adding to his collection. An example of what one of these suggested checklists might look like is included at the end of this chapter.

If this chapter precludes the purchase of a single expensive reproduction, both the author's time and your time will have been well spent.

Example of a Suggested "Original Versus Reproduction" Checklist for Land and Water Customs Daggers

1. The vast majority of reproductions bear the proofmark of the F. W. Höller *Waffenfabrik*. Although the Höller firm did, in fact, manufacture original Customs daggers *(Nr.85, Zolldolch)*, any piece bearing the Höller proofmark should be viewed with suspicion. The proofmarks are very lightly etched on the reproductions.[13]

2. The grip wire on the original daggers is a *single* twisted aluminum or copper wire. A wider band inset grip wire is being used on the reproduction pieces.

3. Early reproductions had the grip grooves and wire running in the opposite diagonal direction from the originals, i.e., on the reproductions the grooves and wire run from the viewer's lower right to upper left.

4. Wooden pegs are used on the reproductions to hold the grip wire intact. The grip wire on an authentic piece is inserted through a small hole at each end of the grip similar to 2nd Model *Luftwaffe* grip wire.

5. Some of the early reproductions have a plastic rather than a metal scabbard base. A magnet will quickly identify this shortcoming.

6. The authentic scabbard is gradually tapered and the lower scabbard fitting does *not* appear to be too large for the scabbard body as with the spurious pieces.

7. The authentic scabbard does *not* reflect the two vertical indentation lines common on the reproductions.

8. The definition of the pebbling on the scabbard leather is less distinct on the originals than on the reproductions.

9. Crudeness of the hilt metal fittings should not be used as a criterion for a spurious dagger. The authentic Customs daggers also reflected a lack of detail when compared to the quality workmanship of other models, such as the similar Army dagger.

10. The early reproduction models are extremely light when compared to the original models. However, this shortcoming was remedied with the Water Customs reproductions when a switch was made to solid brass hilt fittings.

11. Reproductions will usually be encountered without accompanying hangers. Needless to say, however, authentic hangers can be added to a counterfeit dagger.

12. The blades on the reproductions are modern manufacture and not as fine as the original blades, being matte finish instead of high polish.

13. Authentic Water Customs daggers are *extremely* rare and considered to be one of the rarest of all Nazi issue daggers. Thus, any Water Customs dagger should be thoroughly checked out prior to purchasing.

CHAPTER TWENTY-SIX

Solingen Update

"Under the spreading chestnut tree the village smithy stands; the smith a mighty man is he with large and sinewy hands. And the muscles of his brawny arms are strong as iron bands." — Henry Wadsworth Longfellow

During my second career in the field of writing about and buying and selling militaria (after an earlier career of serving twenty-one years as a U.S. Army Airborne Ranger Infantry Officer), I have experienced what I consider to be three tremendous opportunities. The first occurred during the 1960-1970 timeframe when I met and became good friends with the late LTC (Ret.) James Atwood. Jim's exploits in the field of German military are legendary and well documented (see Chapter 2 of *Collecting the Edged Weapons of the Third Reich, Volume II*). Over the course of some 8-10 years, Jim provided me with the opportunity to purchase the "lions share" of all of the thousands of original parts he had acquired in the blade-producing city of Solingen during his historic (to collectors, anyway) 1959-1960 forays into that city when he personally purchased the leftover stock of several of the large edged weapon manufacturing firms. The parts that, indeed, numbered in the thousands also included some invaluable components like a sizeable number of the original diamond-studded 2nd Model Navy Honor dagger pommels Jim had purchased directly from the Eickhorn family. He also very kindly provided me the opportunity to purchase the vast majority of his original 1930s Solingen literature consisting of hundreds of original factory sales catalogs; the priceless, original Paul Casberg artwork; 1930s

The well-known J. A. Henckels *Zwillingswerk* firm is pictured during the 1930s. Note the firm's "twins" logo has been incorporated into the brickwork of the main building in Solingen. This enormous company remains fully operational today and distributes their quality *Besteck* (tableware) worldwide. *Photo by Cramers K.G., Dortmund, Germany.*

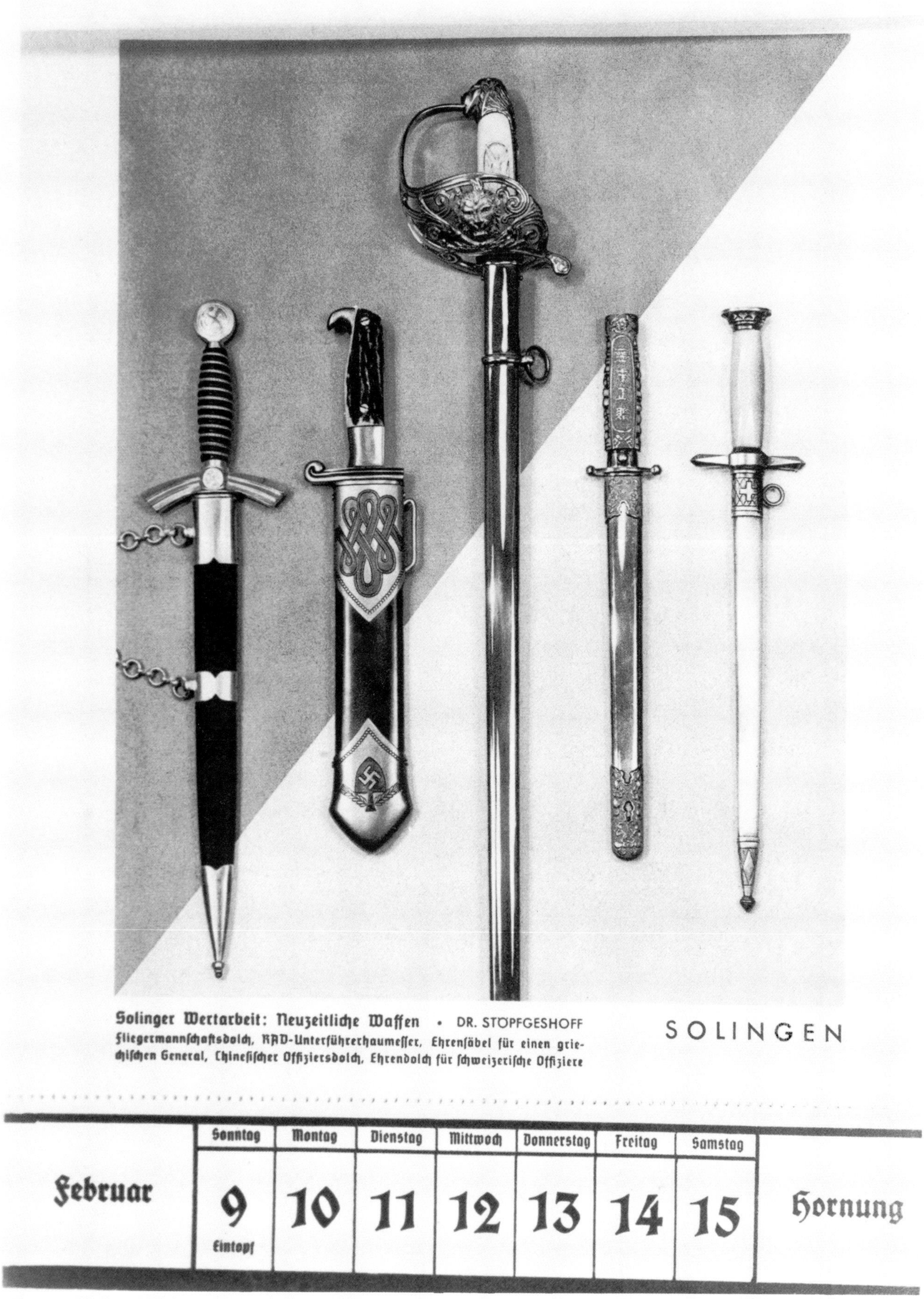

This page from an original 1930s Solingen calendar clearly illustrates the extensive assortment of dress daggers being manufactured in the city during the Third Reich. Included with the 1st Model *Luftwaffe* and *RAD* Enlisted daggers as new *Waffen* (arms), are a Japanese and Swiss dagger. *Johnson Reference Books & Militaria Archives. Photo by LTC (Ret.) Thomas M. Johnson.*

factory photographs; and, literally, hundreds of the original Carl Eickhorn copper and wooden printing clichés used for printing their numerous company sales catalogs, letterheads, and advertisements. A sizeable number of the above historical items have appeared in my own publications on the subject of German edged weaponry and are well known among the numerous members of this hobby.

Another "golden opportunity" occurred much later during August 1989 when I happened to, simply, be in the right place at the right time. Having missed the previous Carl Eickhorn factory bankruptcy sale during August 1976, I was dismayed at the innumerous valuable and irreplaceable original items that most likely ended up in the Solingen *Mülldeponie* (city dump) and lost forever to the collecting community. I vowed to try to be available and on hand, when and if, another Solingen edged weapon firm ever called it quits. In the fall of 1989, the large WKC firm in Solingen had not called it quits, but had experienced a change in ownership with the untimely death of the previous owner, Herr Hans Kolping. I immediately contacted the new owner and was invited to visit the firm that same eventful day.

I was astounded to discover that all of the factory records of the large WKC Firm in Solingen still remained intact. Although this firm, as well as the other existing Solingen manufacturers, claimed for years that all of their records, parts, and machinery, etc. were destroyed during the hostilities of World War II, the opposite proved to be the case with this company, and, I suspect, with several others. Having personally made countless visits to "The City of Swords" since as early as 1960, this came as a total surprise to me! As those readers who know me well realize, finding source research materials on the subject of German edged weapons has always meant more to me than even finding the actual hardware itself; i.e., a previously unseen Solingen factory sales catalog has always held much more fascination for me than even an unusual, original sidearm. Perhaps, what this says is that this author is more of a researcher than he is a collector. Therefore, it is difficult to describe my pleasure and fascination when, during the month of August 1989, the owner of the WKC Firm led me up a narrow set of steel stairways to a small, dark, dank attic on the top floor of the factory. All that was available to illuminate the attic was one very dim, exposed light bulb. Quickly noticing that everything in the attic was covered with a solid inch of black dust that had taken decades to accumulate, made me realize that the opportunity of a researcher's lifetime was being presented to me! On three walls of the attic was a series of bookshelves, and the shelves were completely lined with large, black commercial two-ring German binders maintaining all of the company records from the late 1920s through the 1950s. Also of note to collectors was the fact that there was an ample supply of original Imperial and Third Reich WKC catalogs, including Fencing catalogs, Export catalogs, and a series of 1933-1945 WKC Edged Weapon catalogs. While nearly all collectors are familiar with the wartime WKC catalog featuring a blue cover and silver knight's head logo, I discovered three other previously unseen WKC catalogs that were printed during the Third Reich era. Also discovered were large leftover supplies of original 1933-1945 WKC glass advertising calendars with movable, rotating dates; cardboard point-of-purchase advertising signs; several different price lists; 1930s company letterheads, notepads and mailing envelopes bearing the ubiquitous WKC knight head logo. Lastly, in a large, wooden bin were approximately 100 of the original, metal WKC advertising signs and several bins of original WKC Third Reich dagger and sword parts. Fortunately, I was given the opportunity to purchase the bulk of the contents located in the WKC Factory attic.

Not quite a lost art in Solingen, blade smith Willi Ulrich still polishes sword blades today in his small shop in Gräfrath. This method of utilizing a large rotating grinding wheel has remained essentially the same for centuries with the sole exception of electricity replacing water power to turn the wheel. *Photo by the author.*

Although it will take probably the next few years to ferret through, have translated, and absorb the contents of these black binders, the additional knowledge that will be made available to the serious edged weapon collector is unlimited. Although I have had numerous personal thrills in collecting and researching German edged weapons, nothing that I have experienced in the past could compare with standing in the dark, musty attic of WKC and looking at what I recognized to be an untapped windfall of information on the Solingen blade industry during the 1933-1945 time period. While this information might prove insignificant to some, after having authored, at that time, some twelve books on the subject, I felt like I was standing in my own personal Library of Congress and National Archives combined. Now I had had the opportunity to experience, first-hand, the thrill that James P. Atwood often repeated about his early trips to the Solingen factories and finding literally thousands of leftover parts and hundreds of completed daggers tucked away in musty attics and cellars of various Solingen manufacturing firms.

The third golden opportunity for me was not the result of any "inside track", but, once again, simply a case of, fortunately, being at the right place at the right time. During 1997, the large E.&F. Hörster firm finally realized the inevitable and declared bankruptcy. Needless to say, the demand for quality edged weaponry in huge quantities drastically declined with the demise of Hitler's Third Reich. The entire industry suffered the reduction from each branch of the huge *Wehrmacht* and every *Nazi* political organization wearing their own distinctive edged weapon sidearm to only the German Navy maintaining the tradition post-war. Additionally, in modern times, the extremely high labor costs in Germany made it very difficult for German blade makers to compete with several other countries with their own expanding blade industries. A few of the numerous photographs taken on that memorable day are reproduced in this chapter to document the event. It is, again, interesting to note that although many of the major Solingen edged weapon firms stated for years, in answer to serious collector inquiries from around the world (including many of my own), that "all of our records, weapons, parts, etc. were destroyed during the war", here was, literally, another *entire room* full of leftover parts.

The huge number of available original dagger and sword parts, old literature, calendars, etc. confronted me with the same dilemma facing Jim Atwood in 1960; i.e., where to store boxes and boxes of parts and documents. Fortunately, I, like Atwood before me, was able to arrange for a storage facility in Germany, which I still have available today.

The entrance to the E. & F. Hörster *GmbH* edged weapon factory on *Katternberger Strasse* in the center of Solingen was marked with this brass plate and manned by a security guard. *Photo by the author.*

Included in the above photograph are shelves in the Hörster firm housing some original Hörster manufacturing dies, *Luftwaffe* dagger fittings, and various leftover parts for fencing swords. *Photo by the author.*

While the E. & F. Hörster firm is not an acknowledged manufacturer of the Government Official model dress dagger, these cardboard cartons of leftover Government Official dagger parts may say otherwise. Perhaps many of the Government Official daggers with plain blades (without a manufacturer's trademark) were, in fact, manufactured by the Hörster firm. Some of the other cartons pictured contained hundreds of Third Reich Army dagger unfinished scabbard halves. What would appear to be a room full of piled-up junk to many was viewed in a totally different light by the author. *Photo by the author.*

OPPOSITE
TOP: The author is pictured in 1997 in a small, nondescript room on the second floor of the large E. & F. Hörster factory that was, literally, filled with original leftover Imperial and Third Reich dagger and sword parts. Held are the original Hörster manufacturing dies for the obverse and reverse Army dagger crossguard. *Photo by Hörster employee Frank Schmitz.*

BOTTOM: Leftover, unused leather sword scabbard shells and some Third Reich sword blades are visible in this picture. *Photo by the author.*

ABOVE: E. & F. Hörster shelves of leftover Imperial and Third Reich dagger and sword parts. Note the small carton of original Army dagger hanger parts visible at the lower left. *Photo by the author.*

Frank Schmitz of the E. & F. Hörster firm is pictured displaying a Third Reich Army saber lion head pommel and backstrap. Shelves of original dagger and sword parts can be observed in the background. *Photo by the author.*

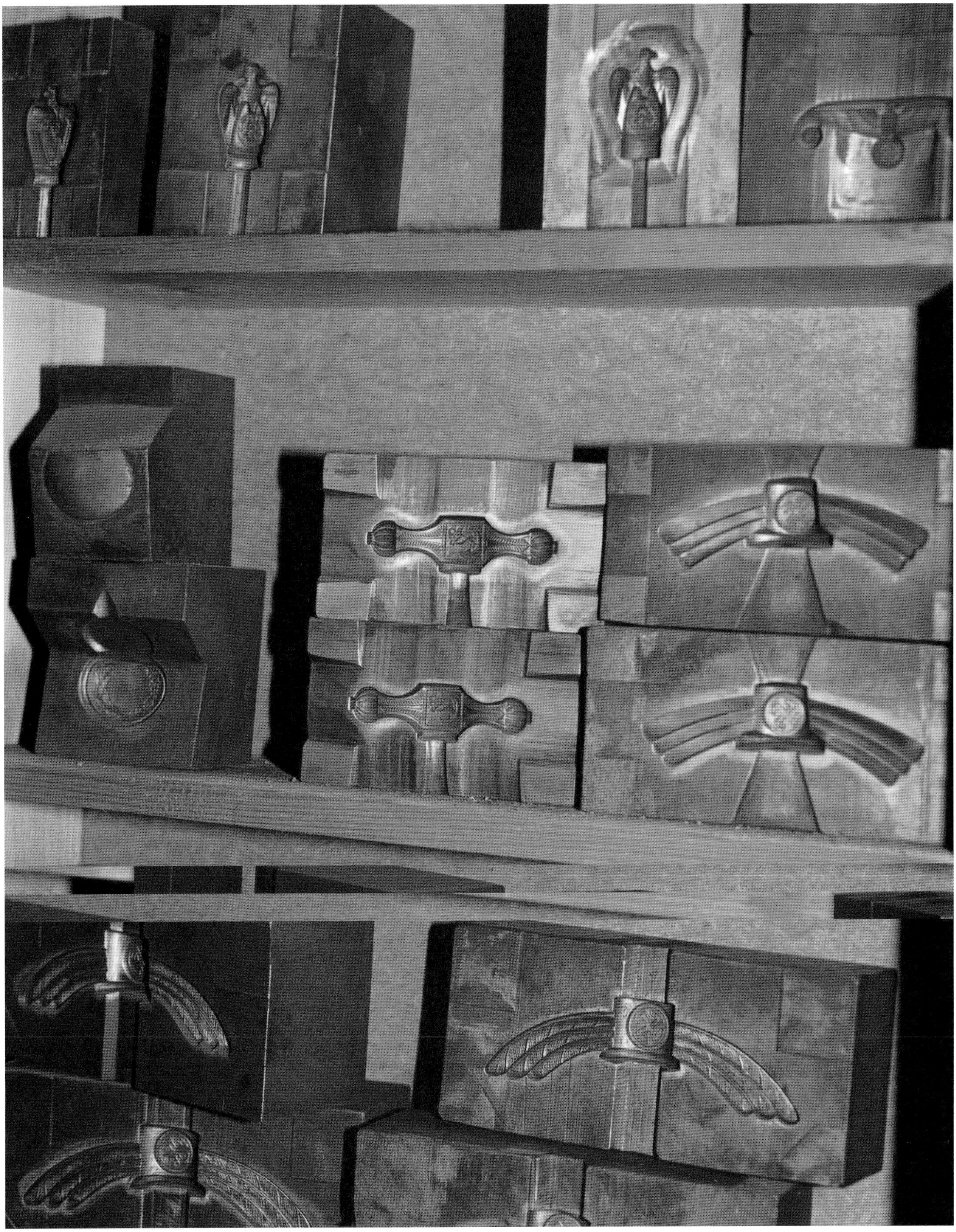

German edged weapon collectors/researchers will certainly recognize the many original Hörster manufacturing dies pictured, i.e., from top to bottom, the 2nd model Navy dagger pommel, the obverse Army dagger crossguard, the 2nd model Navy dagger crossguard, the 1st model dagger *Luftwaffe* crossguard, and the *Luftwaffe* sword crossguard. *Photo by the author.*

Third generation sword maker and former E. & F. Hörster employee Frank Schmitz is pictured at the factory examining a metal container of *Luftwaffe* sword scabbard fittings. Schmitz's grandfather started as a young man in the blade business and worked an entire career of some 38 years. Schmitz's father also worked his total career in the same industry. *Photo by the author.*

The author checks out the final order of E. & F. Hörster U.S. Marine Corps sabers. The company completed this order prior to closing their doors for the final time in 1997. *Photo by Hörster employee Frank Schmitz.*

Third generation sword maker and former E. & F. Hörster employee Frank Schmitz is pictured at his home in Solingen proudly displaying the large handmade Turkish Damascus advertising sword which was proudly displayed on the cover of one of the postwar Hörster firm catalogs. *Photo by the author.*

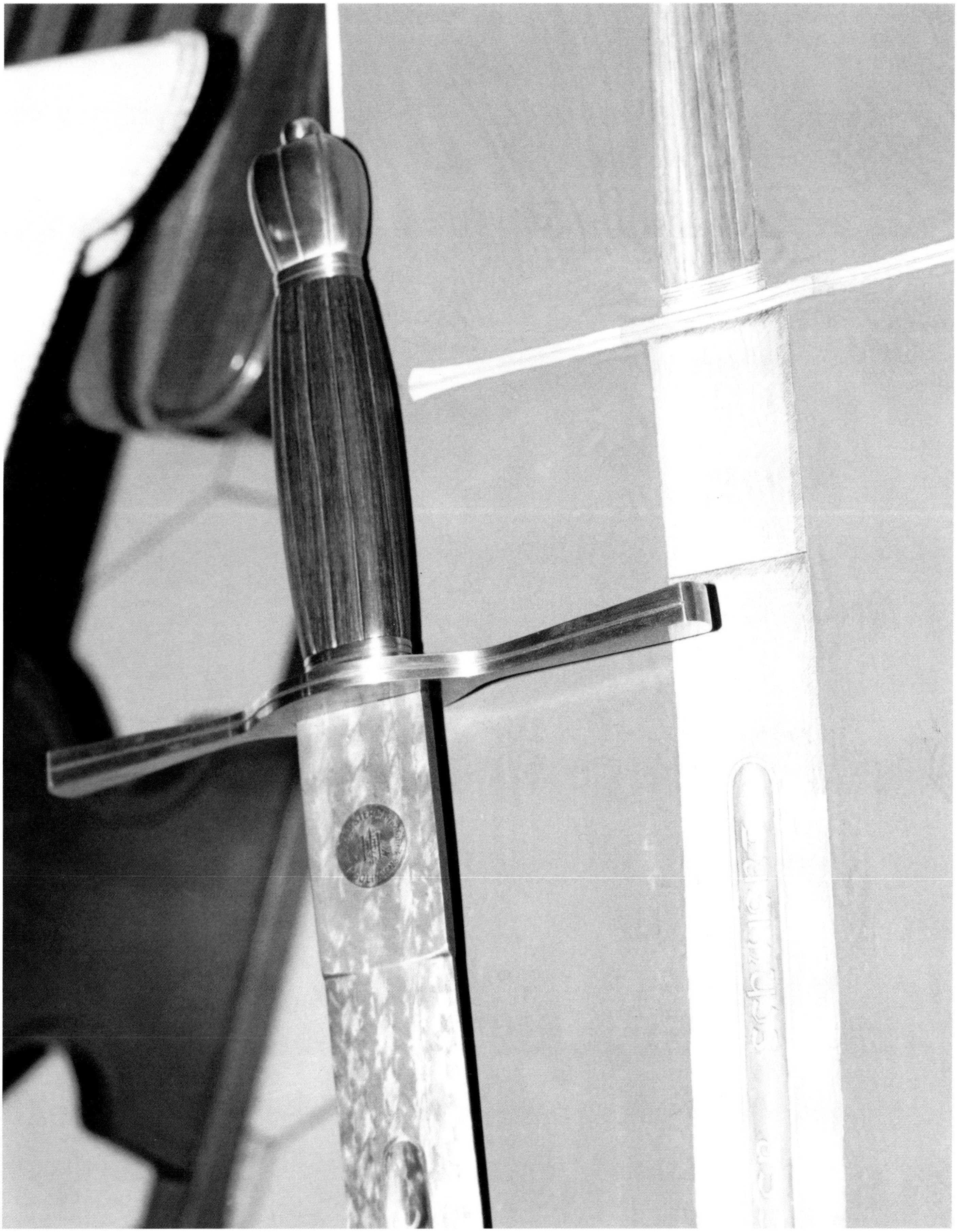

Close-up view of the massive handmade Turkish Damascus advertising sword that was prominently displayed for years in the Hörster factory visitors reception room. This elaborate one-of-a-kind sword also graced the cover of a postwar E. & F. Hörster sales catalog. Note the distinctive five bands of Turkish Damascus steel and the inset circular Hörster logo. *Photo by the author.*

Bibliography

Angolia, John R. *Daggers, Bayonets, & Fighting Knives of Hitler's Germany*. Mountain View, California: R. James Bender Publishing Co., 1971.

Atwood, Major James P. *The Daggers and Edged Weapons of Hitler's Germany*. Berlin, Germany: *Omnium-Druck* and *Verlag*, 1965.

Atwood, Major James P. *World War II Treasure Hunt for Hitler's Daggers of Glory*: SAGA (Vol. 34, No. 2) May 1967, p.8.

Berry, David. *An Introduction to the Daggers of the Third Reich*: The American Blade (Vol. 1, No. 2) July – August 1973, p.22.

Greenwade, Albert ("Skipper"). *Me Fecit Solingen*. Houston, Texas: published privately by the author, 1987.

Hughes, Gordon U. *German Military Fighting Knives*. Brighton, England: The Benedict Press, 1976.

Johnson, Thomas M. *Collecting the Edged Weapons of the Third Reich, Volume I*. Columbia, South Carolina: The R. L. Bryan Co., 1975.

Johnson, Thomas M. *Collecting the Edged Weapons of the Third Reich, Volume II*. Columbia, South Carolina: The R. L. Bryan Co., 1976.

Johnson, Thomas M. *Collecting the Edged Weapons of the Third Reich, Volume III*. Columbia, South Carolina: The R. L. Bryan Co., 1978.

Johnson, Thomas M. *Collecting the Edged Weapons of the Third Reich, Volume IV*. Columbia, South Carolina: The R. L. Bryan Co., 1981.

Johnson, Thomas M. *Collecting the Edged Weapons of the Third Reich, Volume V*. Columbia, South Carolina: The R. L. Bryan Co., 1985.

Johnson, Thomas M. *Collecting the Edged Weapons of the Third Reich, Volume VI*. Columbia, South Carolina: The R. L. Bryan Co., 1992.

Johnson, Thomas M. *Collecting the Edged Weapons of the Third Reich, Volume VII*. Columbia, South Carolina: The R. L. Bryan Co., 1994.

Johnson, Thomas M. *Collecting the Edged Weapons of the Third Reich, Volume VIII*. Columbia, South Carolina: The R. L. Bryan Co., 1996.

Johnson, Thomas M.. *Third Reich Edged Weapon Accouterments*. Columbia, South Carolina: The R. L. Bryan Co., 1978.

Johnson, Thomas M. *Collectors' Handbook of German World War II Daggers*. Columbia, South Carolina: The R. L. Bryan Co., 1979.

Johnson, Thomas M. *The Era of Military Edged Weapons:* The Classic Collector (Vol. 1, No. 1), Fall 1973, p. 34.

Johnson, Thomas M. *World War II German War Booty: A Study in Photographs*. Pennsylvania: Schiffer Publishing Ltd., 2003.

Klietmann, Dr. Kurt-Gerhard. *German Daggers and Dress Sidearms of World War II*. Falls Church, Virginia: Field and Fireside, Inc., 1967.

Kreutz, Heinrich ; Hofmann, Rolf; Johnson, Thomas M.; and Reddick, J. Rex. *Edged Weapon Accouterments of Germany 1800 – 1945*. Pottsboro, Texas: Reddick Enterprises, 2002.

Ormsby, John R., Jr. *Daggers of the Third Reich*. Charlotte, North Carolina: published privately by the author, 1971.

Presentation and Rare Daggers of Hitler's Third Reich: Dowd Private Research Project (United States Military Academy), 1962.

Stephens, F. J. *A Guide to Nazi Daggers, Swords, and Bayonets*. Leigh, Lancs, England: published privately by the author, 1965.

Stephens, F. J. *Edged Weapons of the Third Reich, 1933 – 1945*. London, England: Almark Publishing Col, Ltd., 1972.

Stephens, F. J. *Reproduction? Recognition!*. Leigh, Lancs, England: published privately by the author, 1976.

Venner, Dominique. *Les Armes Blanches du Ill Reich*. Paris, France: Pensee Moderne, 1977.

Weinand, Ronald J. *NPEA Daggers and Associated Knives, A Collector's Guide*. Quincy, Illinois: published privately by the author, 1988.

Wittmann, Thomas T. *Exploring the Dress Daggers of the German Army*. Pennsylvania: Alcom Printing Group, Inc., 1995.

Wittmann, Thomas T. *Exploring the Dress Daggers of the German Luftwaffe*. Pennsylvania: Alcom Printing Group, Inc., 1997.

Wittmann, Thomas T. *Exploring the Dress Daggers of the German Navy*. Pennsylvania: Alcom Printing Group, Inc., 1999.

Wittmann, Thomas T. *Exploring the Dress Daggers of the German SS*. Pennsylvania: Alcom Printing Group, Inc., 2003.

Index